Geometric Measure Theory

A Beginner's Guide

Here as a child I watched my mom blow soap bubbles.
My dad also encouraged all my interests.
This book is dedicated to them with admiration.

Photograph courtesy of the Morgan family;
taken by the author's grandfather, Dr. Charles W. Selemeyer.

SECOND EDITION

Geometric Measure Theory

A Beginner's Guide

Frank Morgan
Department of Mathematics
Williams College
Williamstown, Massachusetts

Illustrated by

James F. Bredt
Department of Mechanical Engineering
Massachusetts Institute of Technology
Cambridge, Massachusetts

ACADEMIC PRESS
San Diego Boston New York
London Sydney Tokyo Toronto

Front cover photograph: John Sullivan's computer rendering of a soup bubble cluster, obtained by area minimization on the Surface Evolver of Ken Brakke of the Geometry Center. See Preface and Chapter 13 for additional information.

Back cover photograph: John Sullivan's computer rendering of an unstable double bubble, never seen in nature, but troublesome to mathematical theory. See Preface and Chapter 13 for additional infomation.

This book is printed on acid-free paper. ♾

Academic Press, Inc.
A Division of Harcourt Brace & Company
525 B Street, Suite 1900, San Diego, California 92101-4495

United Kingdom Edition published by
Academic Press Limited
24-28 Oval Road, London NW1 7DX

Library of Congress Cataloging-in-Publication Data

Morgan, Frank.
Geometric measure theory : a beginner's guide / Frank Morgan : illustrated by James F. Bredt. -- 2nd ed.
p. cm.
Includes bibliographical references (p.) and indexes.
ISBN 0-12-506857-3
1. Geometric measure theory. I. Title.
QA312.M67 1995
515'.41--dc20 94-24913
CIP

PRINTED IN THE UNITED STATES OF AMERICA
95 96 97 98 99 00 BB 9 8 7 6 5 4 3 2 1

Contents

Preface

Singular geometry governs the physical universe: soap bubble clusters meeting along singular curves, black holes, defects in materials, chaotic turbulence, crystal growth. The governing principle is often some kind of energy minimization. Geometric measure theory provides a general framework for understanding such minimal shapes, *a priori* allowing any imaginable singularity and then proving that only certain kinds of structures occur.

Jean Taylor used new tools of geometric measure theory to derive the singular structure of soap bubble clusters and sea creatures, recorded by J. A. F. Plateau over a century ago (see Section 13.9). R. Schoen and S.-T. Yau used minimal surfaces in their original proof of the positive mass conjecture in cosmology. David Hoffman and his collaborators used modern computer technology to discover some of the first new complete embedded minimal surfaces in a hundred years (Figure 6.1.3), some of which look just like certain polymers. Other mathematicians are now investigating singular *dynamics*, such as crystal growth. New software computes crystals growing amidst swirling fluids and temperatures, as well as bubbles in equilibrium, as on the front cover of this book. (See Section 13.12.)

Many of the most basic questions remain open. A single round soap bubble finds the least-area way to enclose a given volume of air. It remains an open mathematical question today whether the familiar *double* bubble actually provides the least-area way to enclose and separate the two given volumes of air; the back cover of this book shows one losing competitor. The analogous planar result for the least-perimeter way to enclose and separate two given areas appeared in 1993 in a paper by a group of undergraduates. (See Chapter 13.)

This little book provides the newcomer or graduate student with an illustrated introduction to geometric measure theory: the basic ideas, terminology, and results. It developed from my one-semester course at MIT for graduate students with a semester of graduate real analysis behind them. I have included a few fundamental arguments and a superficial discussion of the regularity theory, but my goal is merely to introduce the subject and make the standard text, *Geometric Measure Theory* by H. Federer, more accessible.

Other good references include L. Simon's *Lectures on Geometric Measure Theory*, E. Giusti's *Minimal Surfaces and Functions of Bounded Variation*, R. Hardt and Simon's *Seminar on Geometric Measure Theory*, Simon's *Survey Lectures on Minimal Submanifolds*, J. C. C. Nitsche's *Lectures on Minimal Surfaces* (now available in English), R. Osserman's updated *Survey of Minimal Surfaces*, H. B. Lawson's *Lectures on Minimal Submanifolds*, and A. T. Fomenko's new books on *The Plateau Problem*. S. Hildebrandt and A. Tromba offer a beautiful popular gift book for your friends, reviewed by Morgan [10]. J. Brothers assembled a list of open problems. There is an excellent new *Questions and Answers about Area Minimizing Surfaces and Geometric Measure Theory* by F. Almgren [4], who also wrote a review [5] of the first edition of this book.

It was from Fred Almgren, whose geometric perspective this book attempts to capture and share, that I first learned geometric measure theory. I thank many graduate students for their interest and suggestions, especially Benny Cheng, Gary Lawlor, Robert McIntosh, Mohamed Messaoudene, and Marty Ross. I also thank typists Lisa Court, Louis Kevitt, and Marissa Barschdorf. Jim Bredt first illustrated an article of mine as a member of the staff of *Link*, a one-time MIT student newspaper. I feel very fortunate to have him with me again on this book. I am grateful for help with this new edition from many friends, notably Tim Murdoch, Yoshi Giga and his students, who prepared the Japanese translation, and especially John Sullivan. I would like to thank my new editor, Peter Renz, and my original editor and friend Klaus Peters (who with his wife Alice has launched their own company A K Peters). A final thank you goes to all who contributed to this book, at MIT, Rice, Stanford, and Williams. Some support was provided by National Science Foundation grants and by my Cecil and Ida Green Career Development Chair at MIT.

This second edition includes updated material and references, corrections, and a new chapter on soap bubble clusters, including the 1994 counterexample by Weaire and Phelan to the 1887 Kelvin Conjecture. Much new material responds to questions I get about geometric measure theory, regularity results in manifolds, varifolds, $(\mathbf{M}, \varepsilon, \delta)$-minimal sets, calibrations, and so on.

In this edition bibliographic references are simply by author's name, sometimes with an identifying numeral or section reference in brackets. Following a useful practice of Nitsche [2], the bibliography now includes cross-references to each citation.

F. M.
Manhattan, December 9, 1994
Frank.Morgan@williams.edu

CHAPTER 1

Geometric Measure Theory

Geometric measure theory could be described as differential geometry, generalized through measure theory to deal with maps and surfaces that are not necessarily smooth, and applied to the calculus of variations. It dates from the 1960 foundational paper of Herbert Federer and Wendell Fleming on "Normal and Integral Currents," recognized by the 1986 AMS Steele Prize for a paper of fundamental or lasting importance (see Figure 1.0.1). This chapter will give a rough outline of the purpose and basic concepts of geometric measure theory. Later chapters will take up these topics more carefully.

1.1. Archetypical Problem. Given a boundary in $\mathbf{R}^n$, find the surface of least area with that boundary. See Figure 1.1.1. Progress on this problem depends crucially on first finding a good space of surfaces to work in.

1.2. Surfaces as Mappings. Classically, one considered only two-dimensional surfaces, defined as mappings of the disc. See Figure 1.2.1. Excellent references include J. C. C. Nitsche's *Lectures on Minimal Surfaces* [2], now available in English, R. Osserman's updated *Survey of Minimal Surfaces*, and H. B. Lawson's *Lectures on Minimal Submanifolds*. It was not until about 1930 that J. Douglas and T. Rado surmounted substantial inherent difficulties to prove that every smooth Jordan curve bounds a disc of least mapping area. Almost no progress was made for higher-dimensional surfaces (until in a surprising turnaround B. White [1] showed that for higher-dimensional surfaces the geometric measure theory solution actually solves the mapping problem too).

Figure 1.0.1. Wendell Fleming, one of the founders of geometric measure theory. Photograph courtesy of W. Fleming.

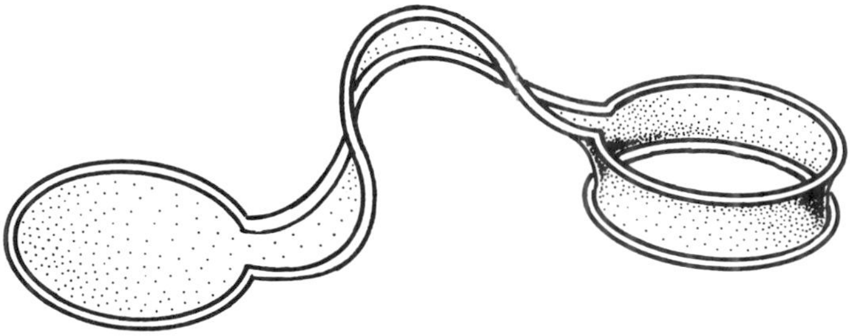

Figure 1.1.1. The surface of least area bounded by two given Jordan curves.

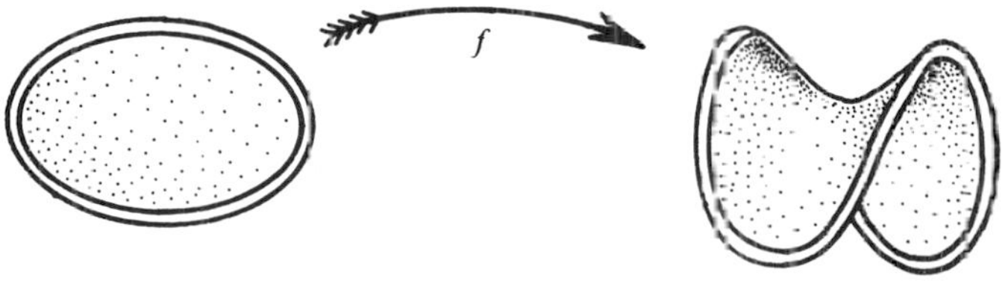

Figure 1.2.1. Surface realized as a mapping, f, of the disc.

Along with its successes and advantages, the definition of a surface as a mapping has certain drawbacks:

(1) There is an inevitable *a priori* restriction on the types of singularities that can occur;
(2) There is an *a priori* restriction on the topological complexity; and
(3) The natural topology lacks compactness properties.

The importance of compactness properties appears in the direct method described in the next section.

1.3. The Direct Method. The direct method for finding a surface of least area with a given boundary has three steps.

(1) Take a sequence of surfaces with areas decreasing to the infimum.
(2) Extract a convergent subsequence.
(3) Show that the limit surface is the desired surface of least area.

Figures 1.3.1–1.3.4 show how this method breaks down for lack of compactness in the space of surfaces as mappings, even when the given bound-

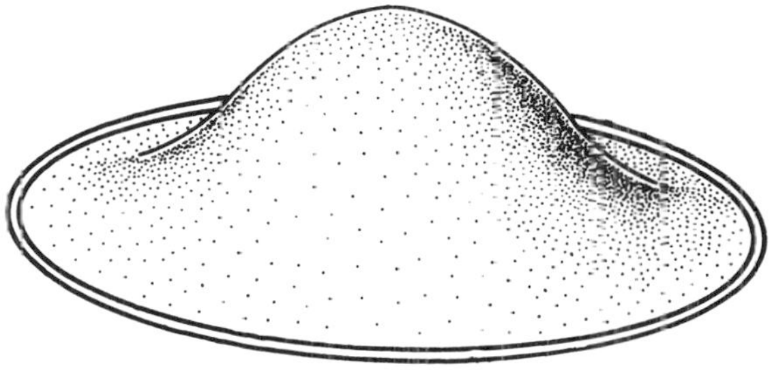

Figure 1.3.1. A surface with area $\pi + 1$.

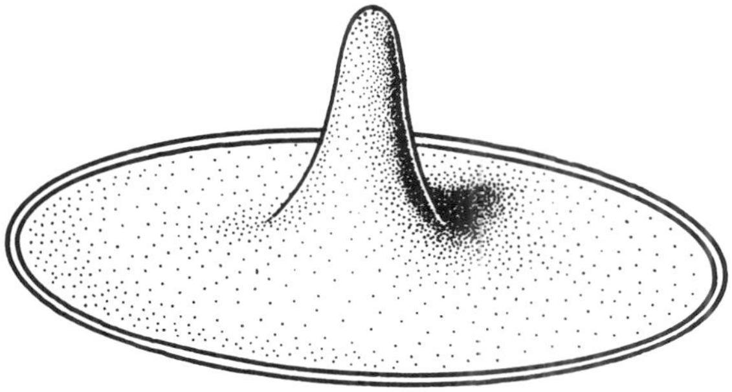

Figure 1.3.2. A surface with area $\pi + \frac{1}{4}$.

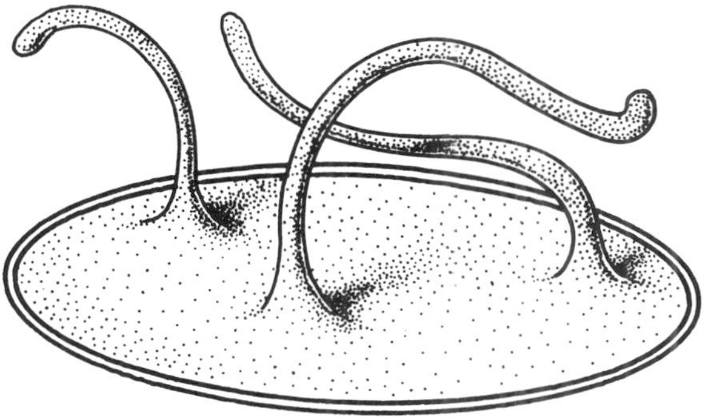

Figure 1.3.3. A surface with area $\pi + \frac{1}{16}$.

ary is the unit circle. By sending out thin tentacles toward every rational point, the sequence could include all of $\mathbf{R}^3$ in its closure!

1.4. Rectifiable Currents. An alternative to surfaces as mappings is provided by *rectifiable currents*, the m-dimensional, oriented surfaces of geometric measure theory. The relevant functions f: $\mathbf{R}^m \to \mathbf{R}^n$ need not be smooth, but merely *Lipschitz*, i.e.,

$$|f(x) - f(y)| \leq C|x - y|,$$

for some "Lipschitz constant" C.

Fortunately there is a good m-dimensional measure on $\mathbf{R}^n$, called *Hausdorff measure*, $\mathscr{H}^m$. Hausdorff measure agrees with the classical mapping area of an embedded manifold, but it is defined for all subsets of $\mathbf{R}^n$.

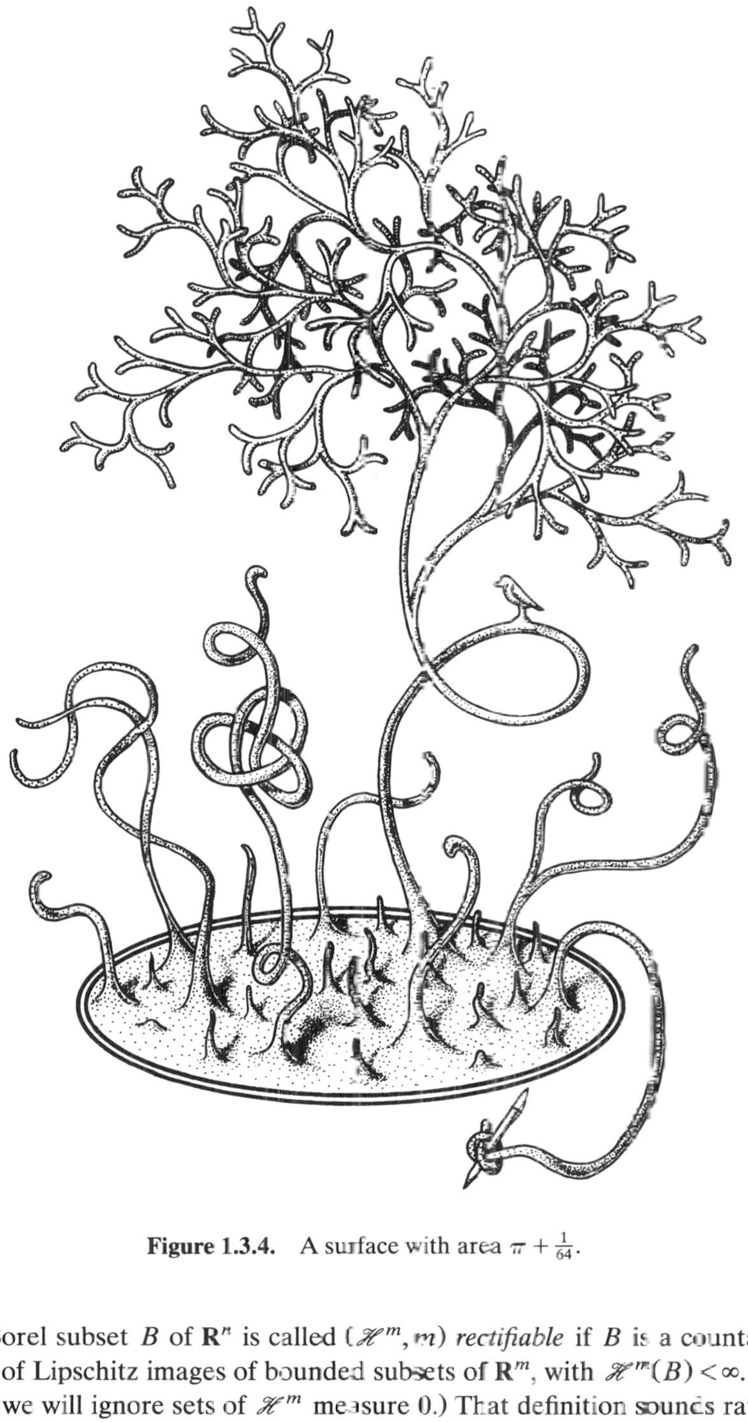

Figure 1.3.4. A surface with area $\pi + \frac{1}{64}$.

A Borel subset B of $\mathbf{R}^n$ is called $(\mathscr{H}^m, m)$ *rectifiable* if B is a countable union of Lipschitz images of bounded subsets of $\mathbf{R}^m$, with $\mathscr{H}^m(B) < \infty$. (As usual, we will ignore sets of $\mathscr{H}^m$ measure 0.) That definition sounds rather general, and it includes just about any "m-dimensional surface" I can imagine. Nevertheless, these sets will support a kind of differential geometry; for example, it turns out that a rectifiable set B has a canonical tangent plane at almost every point.

Finally, a *rectifiable current* is an oriented rectifiable set with integer multiplicities, finite area, and compact support. By general measure theory, one can integrate a smooth differential form φ over such an oriented rectifiable set S, and hence view S as a *current*, i.e., a linear functional on differential forms,

$$\varphi \mapsto \int_S \varphi.$$

This perspective yields a new natural topology on the space of surfaces, dual to an appropriate topology on differential forms. This topology has useful compactness properties, given by the fundamental Compactness Theorem below. Viewing rectifiable sets as currents also provides a boundary operator ∂ from m-dimensional rectifiable currents to $(m-1)$-dimensional currents, defined by

$$(\partial S)(\varphi) = S(d\varphi),$$

where $d\varphi$ is the exterior derivative of φ. By Stokes's Theorem, this definition coincides with the usual notion of boundary for smooth, compact manifolds with boundary. In general, the current ∂S is not rectifiable, even if S is rectifiable.

1.5. The Compactness Theorem. *Let c be a positive constant. Then the set of all m-dimensional rectifiable currents T in a fixed large closed ball in $\mathbf{R}^n$, such that the boundary ∂T is also rectifiable and such that the area of both T and ∂T are bounded by c, is compact in an appropriate weak topology.*

1.6. Advantages of Rectifiable Currents. Notice that rectifiable currents have none of the three drawbacks mentioned in Section 1.2. There is certainly no restriction on singularities or topological complexity. Moreover, the compactness theorem provides the ideal compactness properties. In fact, the direct method described in Section 1.3 succeeds in the context of rectifiable currents. In the figures of Section 1.3, the amount of area in the tentacles goes to 0. Therefore, they disappear in the limit in the new topology. What remains is the disc, the desired solution.

All of these results hold in all dimensions and codimensions.

1.7. The Regularity of Area-Minimizing Rectifiable Currents. One serious suspicion hangs over this new space of surfaces: The solutions they provide to the problem of least area, the so-called area-minimizing rectifiable currents, may be generalized objects without any geometric significance. The following interior regularity results allay such concerns. (We give more precise statements in Chapter 8.)

(1) A two-dimensional area-minimizing rectifiable current in $\mathbf{R}^3$ is a smooth embedded manifold.
(2) For $m \leq 6$, an m-dimensional area-minimizing rectifiable current in $\mathbf{R}^{m+1}$ is a smooth embedded manifold.

Thus in low dimensions the area-minimizing hypersurfaces provided by geometric measure theory actually turn out to be smooth embedded manifolds. However, in higher dimensions, singularities occur, for geometric and not merely technical reasons (see Section 10.7). Despite marked progress, understanding such singularities remains a tremendous challenge.

CHAPTER 2

Measures

This chapter lays the measure-theoretic foundation, including the definition of Hausdorff measure and covering theory. The general reference is Federer [1, Chapter II].

2.1. Definitions. For us a *measure* μ on $\mathbf{R}^n$ will be what is sometimes called an outer measure: a nonnegative function μ on *all* subsets of $\mathbf{R}^n$ (with the value $+\infty$ allowed, of course), which is *countably subadditive*, i.e., if A is contained in a countable union, $\bigcup A_i$, then

$$\mu(A) \leq \Sigma \mu(A_i).$$

A set $A \subset \mathbf{R}^n$ is called *measurable* if, for all $E \subset \mathbf{R}^n$, $\mu(E \cap A) + \mu(E \cap A^C) = \mu(E)$. The class of measurable sets is a σ-algebra, i.e., closed under complementation, countable union, and countable intersection. If A is a countable disjoint union of measurable sets A_i, then $\mu(A) = \Sigma \mu(A_i)$.

The smallest σ-algebra containing all open sets is the collection of *Borel* sets. A measure μ is called *Borel regular* if Borel sets are measurable and every subset of $\mathbf{R}^n$ is contained in a Borel set of the same measure.

Suppose that μ is Borel regular, A is measurable, and $\varepsilon > 0$. If $\mu(A) < \infty$, then A contains a closed subset C with $\mu(A - C) < \varepsilon$. If A can be covered by countably many open sets of finite measure, then A is contained in an open set W with $\mu(W - A) < \varepsilon$ [Federer, 2.2.3].

All Borel sets are measurable if and only if Caratheodory's criterion holds:

(1) Whenever A_1, A_2 are sets a positive distance apart, then

$$\mu(A_1 \cup A_2) = \mu(A_1) + \mu(A_2).$$

2.2. Lebesgue Measure. There is a unique Borel regular, translation invariant measure on $\mathbf{R}^n$ such that the measure of the unit cube $[0,1]^n$ is 1. This measure is called *Lebesgue measure*, $\mathscr{L}^n$.

2.3. Hausdorff Measure [Federer, 2.10]. Unfortunately, for general "m-dimensional" subsets of $\mathbf{R}^n$ (for $m < n$), it is more difficult to assign an m-dimensional measure. The m-dimensional area of a C^1 map f from a domain $D \subset \mathbf{R}^m$ into $\mathbf{R}^n$ is classically defined as the integral of the Jacobian $J_m f$ over D. [Computationally, at each point $x \in D$, $(J_m f)^2$ equals the sum of the squares of the determinants of the $m \times m$ submatrices of $Df(x)$ or, equivalently, the determinant of $(Df(x))^t Df(x)$.] The area of an m-dimensional submanifold M of $\mathbf{R}^n$ is then defined by calculating it on parameterized portions of M and proving that the area is independent of choice of parameterization.

In 1918, F. Hausdorff introduced an m-dimensional measure in $\mathbf{R}^n$ which gives the same area for submanifolds, but is defined on all subsets of $\mathbf{R}^n$. When $m = n$, it turns out to be equal Lebesgue measure.

Definitions. For any subset S of $\mathbf{R}^n$, define the *diameter* of S

$$\operatorname{diam}(S) = \sup\{|x - y| : x, y \in S\}.$$

Let α_m denote the Lebesgue measure of the closed unit ball $\mathbf{B}^m(\mathbf{0},1) \subset \mathbf{R}^m$. For $A \subset \mathbf{R}^n$, we define the m-dimensional Hausdorff measure $\mathscr{H}^m(A)$ by the following process. For small δ, cover A efficiently by countably many sets S_j with $\operatorname{diam}(S_j) \le \delta$, add up all the $\alpha_m(\operatorname{diam}(S_j)/2)^m$, and take the limit as $\delta \to 0$:

$$\mathscr{H}^m(A) = \lim_{\delta \to 0} \inf_{\substack{A \subset \bigcup S_j \\ \operatorname{diam}(S_j) \le \delta}} \sum \alpha_m \left(\frac{\operatorname{diam}(S_j)}{2} \right)^m.$$

The infimum is taken over all countable coverings $\{S_j\}$ of A whose members have diameter at most δ. As δ decreases, the more restricted infimum cannot decrease, and hence the limit exists, with $0 \le \mathscr{H}^m(A) \le \infty$. In Figure 2.3.1, the two-dimensional area is approximated by $\sum \pi r^2$. The spiral of Figure 2.3.2 illustrates one reason for taking the limit as $\delta \to 0$, since otherwise a spiral of great length could be covered by a single ball of radius 1.

Countable subadditivity follows immediately from the definition. The measurability of Borel sets follows easily from Caratheodory's criterion 2.1(1).

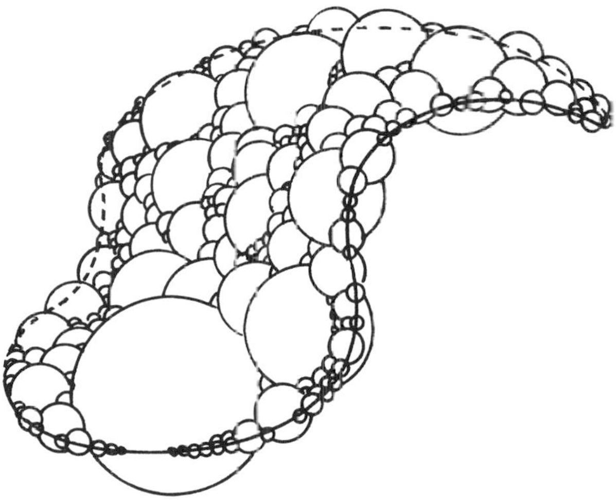

Figure 2.3.1. The Hausdorff measure (area) of a piece of surface A is approximated by the cross-sections of little balls which cover it.

To see that each $A \subset \mathbf{R}^n$ is contained in a Borel set B of the same measure, note first that each S_j occurring in the definition of $\mathscr{H}^m(A)$ may be replaced by its closure, so that $\cup S_j$ is Borel. If $\{S_j^{(k)}\}$ is a countable sequence of coverings defining $\mathscr{H}^m(A)$, then $B = \cap_k \cup_j S_j^{(k)}$ gives the de-

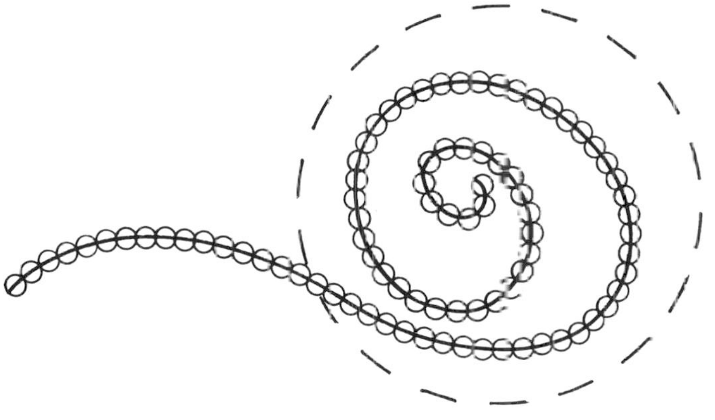

Figure 2.3.2. One must cover by *small* sets to compute length accurately. Here the length of the spiral is well estimated by the sum of the diameters of the tiny balls, but grossly underestimated by the diameter of the huge ball.

sired Borel set. Therefore, $\mathscr{H}^m$ is Borel regular. Later it will be proved that $\mathscr{H}^m$ gives the "correct" area for C^1 submanifolds of $\mathbf{R}^n$.

The definition of Hausdorff measure extends to any nonnegative real dimension. [The definition of α_m is extended by the Γ function: $\alpha_m = \pi^{m/2}/\Gamma(m/2+1)$]. Notice that $\mathscr{H}^0$ is counting measure; $\mathscr{H}^0(A)$ is the number of elements of A.

The *Hausdorff dimension* of a nonempty set A is defined as

$$\begin{aligned}\inf\{m \geq 0: \mathscr{H}^m(A) < \infty\} &= \inf\{m: \mathscr{H}^m(A) = 0\}\\ &= \sup\{m: \mathscr{H}^m(A) > 0\}\\ &= \sup\{m: \mathscr{H}^m(A) = \infty\}.\end{aligned}$$

The equivalence of these conditions follows from the fact that if $m < k$ and $\mathscr{H}^m(A) < \infty$, then $\mathscr{H}^k(A) = 0$ (Exercise 2.4). The Cantor set of Exercise 2.6 turns out to have Hausdorff dimension $\ln 2/\ln 3$. Figure 2.3.3 pictures a

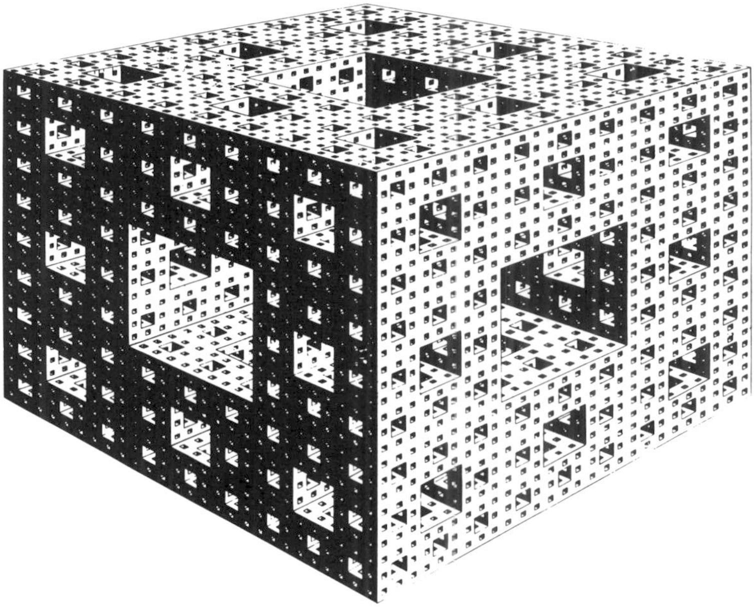

Figure 2.3.3. The Sierpinski sponge is an example of a fractional dimensional set. Its Hausdorff dimension is about 2.7. (From *Studies in Geometry* by Leonard M. Blumenthal and Karl Menger. Copyright 1979 by W. H. Freeman and Company. Reprinted with permission.)

Cantor-like set in $\mathbf{R}^3$, called the Sierpinski sponge, which has Hausdorff dimension of about 2.7.

These Cantor-like sets are self-similar in the sense that certain homothetic expansions of such a set are locally identical to the original set. Self-similarity appears in the coastline of Great Britain and in the mass in the universe. B. Mandelbrot has modeled many natural phenomena by random fractional dimensional sets and processes, called *fractals*. His books, *Fractals* and *The Fractal Geometry of Nature*, contain beautiful, computer-generated pictures of hypothetical clouds, landscapes, and other phenomena.

2.4. Integralgeometric Measure. In 1932, J. Favard defined another m-dimensional measure on $\mathbf{R}^n$ $(m = 0, 1, \ldots, n)$, now called integral-geometric measure, $\mathscr{I}^m$. It turns out that $\mathscr{I}^m$ agrees with $\mathscr{H}^m$ on all smooth m-dimensional submanifolds and other nice sets, but disagrees and often is zero on Cantor-like sets.

Roughly, to define $\mathscr{I}^m(A)$, project A onto an m-dimensional subspace of $\mathbf{R}^n$, take the Lebesgue measure (counting multiplicities), and average over all such projections.

More precisely, let $\mathbf{O}^*(n, m)$ denote the set of orthogonal projections p of $\mathbf{R}^n$ onto m-dimensional subspaces. For general reasons there is a unique measure on $\mathbf{O}^*(n, m)$, invariant under Euclidean motions on $\mathbf{R}^n$, normalized to have total measure 1. For example, the set $\mathbf{O}^*(2, 1)$ of orthogonal projections onto lines through 0 in the plane may be parameterized by $0 \leq \theta < \pi$, and the unique measure is $(1/\pi)\,d\theta$. For $y \in \text{image } p \cong \mathbf{R}^m$, let the "multiplicity function," $N(p|A, y)$, denote the number of points in $A \cap p^{-1}(y)$. Define a normalizing constant,

$$\beta(n,m) = \Gamma\left(\frac{m+1}{2}\right)\Gamma\left(\frac{n-m+1}{2}\right)\Gamma\left(\frac{n+1}{2}\right)^{-1}\pi^{(-1/2)}.$$

Now define the integralgeometric measure of any Borel set B by

$$\mathscr{I}^m(B) = \frac{1}{\beta(n,m)}\int_{p \in \mathbf{O}^*(n,m)}\int_{y \in \text{image } p \cong \mathbf{R}^m} N(p|B, y)\, d\mathscr{L}^m y\, dp.$$

One checks that the function $N(p|B, y)$ is indeed measurable and that $\mathscr{I}^m$ is countably subadditive. Finally extend $\mathscr{I}^m$ to a Borel regular measure by defining for any set $A \subset \mathbf{R}^n$,

$$\mathscr{I}^m(A) = \inf\{\mathscr{I}^m(B) : A \subset B,\ B \text{ Borel}\}.$$

2.5. Densities [Federer, 2.9.12, 2.10.19]. Let A be a subset of $\mathbf{R}^n$. For $1 \le m \le n$, $a \in \mathbf{R}^n$, we define the m-dimensional *density* $\Theta^m(A,a)$ of A at a by the formula

$$\Theta^m(A,a) = \lim_{r \to 0} \frac{\mathscr{H}^m(A \cap \mathbf{B}^n(a,r))}{\alpha_m r^m},$$

where α_m is the measure of the closed unit ball $\mathbf{B}^m(\mathbf{0},1)$ in $\mathbf{R}^m$. For example, the cone

$$C = \{x^2 + y^2 = z^2\}$$

of Figure 2.5.1 has two-dimensional density

$$\Theta^2(C,a) = \begin{cases} 1 & \text{for} \quad a \in C - \{\mathbf{0}\}, \\ 0 & \text{for} \quad a \notin C, \\ \sqrt{2} & \text{for} \quad a = \mathbf{0}. \end{cases}$$

Figure 2.5.1. The cone $\{x^2 + y^2 = z^2\}$ has density 1 everywhere except at the vertex, where it has density $\sqrt{2}$.

Similarly, for μ a measure on $\mathbf{R}^n$, $1 \le m \le n$, $a \in \mathbf{R}^n$, define the m-dimensional density $\Theta^m(\mu, a)$ of μ at a by

$$\Theta^m(\mu, a) = \lim_{r \to 0} \frac{\mu(\mathbf{B}^n(a,r))}{\alpha_m r^m}.$$

Note that for any subset A of $\mathbf{R}^n$, $\Theta^m(A, a) = \Theta^m(\mathscr{H}^m \llcorner A, a)$, where $\mathscr{H}^m \llcorner A$ is the measure defined by

$$(\mathscr{H}^m \llcorner A)(E) \equiv \mathscr{H}^m(A \cap E).$$

Hence density of measures actually generalizes the notion of density of sets.

2.6. Approximate Limits [Federer, 2.9.12]. Let $A \subset \mathbf{R}^m$. A function $f\colon A \to \mathbf{R}^n$ has *approximate limit* y at a if for every $\varepsilon > 0$, $\mathbf{R}^m - \{x \in A\colon |f(x) - y| < \varepsilon\}$ has m-dimensional density 0 at a. We write $y = ap\lim_{x \to a} f(x)$. Note that in particular A must have density 1 at a.

PROPOSITION. *A function $f\colon A \to \mathbf{R}^n$ has an approximate limit y at a if and only if there is a set $B \subset A$ such that B^c has m-dimensional density 0 at a and $f|B$ has the limit y at a.*

Remark. In general, the word *approximate* means "except for a set of density 0."

Proof. The condition is clearly sufficient. To prove necessity, assume that f has an approximate limit y at a. For convenience we assume $y = 0$. Then for any positive integer i,

$$A_i \equiv \mathbf{R}^m - \{x \in A\colon |f(x)| < 1/i\}$$

has density 0 at a. Choose $r_1 > r_2 > \dots$ such that

$$\frac{\mathscr{H}^m(A_i \cap \mathbf{B}^n(a,r))}{\alpha_m r^m} \le 2^{-i}$$

whenever $0 < r \le r_i$. Notice that $A_1 \subset A_2 \subset \dots$. Let $B^C = \bigcup(A_i \cap \mathbf{B}(a, r_i))$.

Clearly $f|B$ has the limit y at a. To show that B^C had density 0 at a, let $r_i > s > r_{i+1}$. Then

$$\begin{aligned}\mathscr{H}^m(B^C \cap \mathbf{B}(a,s)) &\le \mathscr{H}^m(A_i \cap \mathbf{B}(a,s)) + \mathscr{H}^m(A_{i+1} \cap \mathbf{B}(a, r_{i-1})) \\ &\quad + \mathscr{H}^m(A_{i+2} \cap \mathbf{B}(a, r_{i+2})) + \cdots \\ &\le \alpha_m(s^m \cdot 2^{-i} + r_{i+1}^m \cdot 2^{-(i+1)} + r_{i+2}^m \cdot 2^{-(i+2)} + \cdots) \\ &\le \alpha_m \cdot s^m \cdot 2^{-(i-1)}.\end{aligned}$$

Therefore B^C has density 0 at a, as desired.

Definitions. Let $a \in A \subset \mathbf{R}^m$. A function $f\colon A \to \mathbf{R}^n$ is *approximately continuous* at a if $f(a) = \operatorname{ap}\lim_{x \to a} f(x)$. The point a is a *Lebesgue point* of f if $\Theta^m(A^C, a) = 0$ and

$$\frac{1}{\alpha_m r^m} \int_{A \cap \mathbf{B}(a,r)} |f(x) - f(a)| \, d\mathscr{L}^m x \xrightarrow[r \to 0]{} 0.$$

The function f is *approximately differentiable* at a if there is a linear function $L\colon \mathbf{R}^m \to \mathbf{R}^n$ such that

$$\operatorname{ap}\lim_{x \to a} \frac{|f(x) - f(a) - L(x-a)|}{|x - a|} = 0.$$

We write $L = \operatorname{ap} Df(a)$.

The following covering theorem of Besicovitch proves more powerful in practice than more familiar ones, such as Vitali's. It applies to any finite Borel measure φ.

2.7. Besicovitch Covering Theorem [Federer, 2.8.15; Besicovitch]. *Suppose φ is a Borel measure on $\mathbf{R}^n$, $A \subset \mathbf{R}^n$, $\varphi(A) < \infty$, F is a collection of nontrivial closed balls, and* $\inf\{r\colon \mathbf{B}(a,r) \in F\} = 0$ *for all $a \in A$. Then there is a (countable) disjoint subcollection of F that covers φ almost all of A.*

Partial Proof. We may assume that all balls in F have radius at most 1.

Part 1. *There is a constant $\zeta(n)$ such that, given a closed ball, B, of radius r and a collection, C, of closed balls of a radius of at least r which intersect B and which do not contain each other's centers, then the cardinality of C is at most $\zeta(n)$.* This statement is geometrically obvious, and we omit the proof. E. R. Riefenberg [4] proved that for $n = 2$, the sharp bound is 18. (See Figure 2.7.1.)

Part 2. *$\zeta + 1$ subcollections of disjoint balls cover A.* To prove this statement, we will arrange the balls of F in rows of disjoint balls, starting with the largest and proceeding in order of size. (Of course, there may not always be a "largest ball," and actually one chooses a nearly largest ball. This technical point propagates minor corrections throughout the proof, which we will ignore.)

Place the largest ball B_1 in the first row. (See Figure 2.7.2.) Throw away all balls whose centers are covered by B_1.

Figure 2.7.1. At most, 18 larger discs can intersect the unit disc in $\mathbf{R}^2$ without containing each other's centers. Figure courtesy of J. Sullivan.

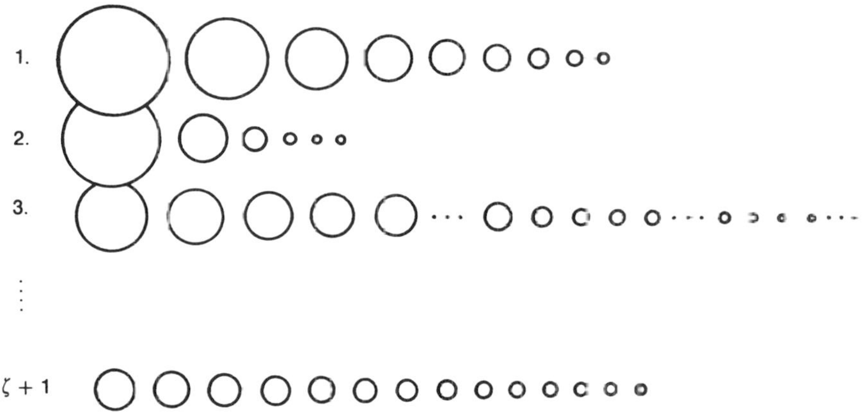

Figure 2.7.2. In the proof of the Besicovitch covering Theorem 2.7, the balls covering A are arranged by size in rows or discarded. Intersecting balls must go in different rows. For the case of $\mathbf{R}^2$, this requires at most $\zeta(2)+1 \leq 19$ rows. Then some row must provide a disjoint cover of at least $1/19$ the total measure.

Take the next largest ball, B_2. If B_2 is disjoint from B_1, place B_2 in the first row. If not, place B_2 in the second row. Throw away all balls whose centers are covered by B_2.

At the nth step, place B_n in the earliest row that keeps all balls in each row disjoint. Throw away all balls whose centers are covered by B_n.

Proceed by transfinite induction. The whole list certainly covers A, since we throw away only balls whose centers are already covered. Each row consists of disjoint balls, by construction. Hence it suffices to show that there are at most $\zeta+1$ nonempty rows. Assume some ball, B, gets put in the $\zeta+2$ row. Then there are balls $D_1,\ldots,D_{\zeta+1}$, at least as large as B already in the first $\zeta+1$ rows and not disjoint from B. No D_j can contain another's center, or the smaller would have been thrown away when the larger was put in. This contradiction of *Part 1* completes the proof of *Part 2*.

PART 3. *Completion of proof.* By *Part 2*, some disjoint subcollection covers $1/(\zeta+1)$ the φ measure of A. Hence some finite disjoint subcollection covers a closed subset $A_1 \subset A$ with

$$\frac{\varphi(A_1)}{\varphi(A)} \geq \frac{1}{\zeta+2}, \quad \text{i.e., } 1-\frac{\varphi(A_1)}{\varphi(A)} \leq \delta < 1.$$

Repeat the whole process on $A - A_1$ with the balls contained in $A - A_1$ to obtain a finite disjoint subcollection covering $A_2 \subset A$ with

$$1-\frac{\varphi(A_2)}{\varphi(A)} \leq \delta^2.$$

Countably many such repetitions finally yield a countable disjoint subcollection covering φ almost all of A.

We now give three corollaries as examples of the usefulness of Besicovitch's covering theorem.

2.8. Corollary. $\mathscr{H}^n = \mathscr{L}^n$ on $\mathbf{R}^n$.

Proof. We will need the so-called isodiametric inequality, which says that among all sets of fixed diameter, the ball has the largest volume. In other words, for any set S in $\mathbf{R}^n$,

$$\mathscr{L}^n(S) \leq \alpha_n \left(\frac{\operatorname{diam} S}{2}\right)^n.$$

It follows immediately that $\mathscr{H}^n \geq \mathscr{L}^n$.

There happens to be an easy proof of the isodiametric inequality. We may assume that S is symmetric with respect to each coordinate axis, since replacing each intersection of S with a line parallel to the axis by a symmetric interval of the same one-dimensional measure does not change the Lebesgue measure and can only decrease the diameter. But now S is symmetric with respect to the origin and hence is contained in the ball B of the same diameter. Therefore

$$\mathscr{L}^n(S) \le \mathscr{L}^n(B) = \alpha_n \left(\frac{\operatorname{diam} S}{2} \right)^n,$$

as desired. Notice that the symmetrization step is necessary, because an equilateral triangle, for example, is not contained in a ball of the same diameter.

To prove that $\mathscr{H}^n \le \mathscr{L}^n$, we will use the Besicovitch Covering Theorem. First, we note that it suffices to prove that $\mathscr{H}^n(A) \le \mathscr{L}^n(A)$ for A Borel and bounded, or hence for A equal to the open R-ball $\mathbf{U}^n(\mathbf{0}, R) \subset \mathbf{R}^n$, or hence for $A = \mathbf{U}^n(\mathbf{0}, 1)$. An easy computation shows that $\mathscr{H}^n(A) < \infty$. Given $\varepsilon > 0$, choose $\delta > 0$ such that

$$(1) \quad \mathscr{H}^n(A) \le \inf\left\{ \Sigma \alpha_n \cdot \left(\frac{\operatorname{diam} S_i}{2} \right)^n : A \subset \bigcup S_i,\ \operatorname{diam} S_i \le \delta \right\} + \varepsilon.$$

Apply the covering theorem with

$$F = \{\text{closed balls contained in } A \text{ with diameter} \le \delta\}$$

to obtain a disjoint covering G of $B \subset A$ with $\mathscr{H}^n(A - B) = 0$. Let G' be a covering by balls of diameter at most δ of $A - B$ with

$$\sum_{S \in G'} \alpha_n \left(\frac{\operatorname{diam} S}{2} \right)^n \le \varepsilon.$$

Then $G \cup G'$ covers A, and therefore

$$\begin{aligned}
\mathscr{H}^n(A) &\le \sum_{S \in G \cup G'} \alpha_n \left(\frac{\operatorname{diam} S}{2} \right)^n + \varepsilon \\
&\le \sum_{S \in G} \mathscr{L}^n(S) + \sum_{S \in G'} \alpha_n \left(\frac{\operatorname{diam} S}{2} \right)^n + \varepsilon \\
&\le \mathscr{L}^n(A) + \varepsilon + \varepsilon.
\end{aligned}$$

The corollary is proved. The fussing with $A - B$ at the end was necessary because (1) does not apply to B.

2.9. Corollary. *If $A \subset R^n$ is Lebesgue measurable, then the density $\Theta^n(A, x)$ equals the characteristic function $\chi_A(x)$ almost everywhere.*

Proof. It suffices to show that for every measurable set A, $\Theta(A, x) = 1$ at almost all points $x \in A$. (Considering A^C then implies $\Theta(A, x) = 0$ at almost all $x \notin A$.) Assume not. We may assume $0 < \mathscr{L}^n(A) < \infty$. We may further assume that for some $\delta < 1$

$$(1) \qquad \Theta_*(A, a) = \underline{\lim} \frac{\mathscr{L}^n(A \cap \mathbf{B}(a, r))}{\alpha_n r^n} < \delta \quad \text{for all } a \in A,$$

by first choosing δ such that

$$\mathscr{L}^n\{a \in A: \Theta_*(A, a) < \delta\} > 0$$

and then replacing A by $\{a \in A: \Theta_*(A, a) < \delta\}$. Choose an open set $U \supset A$ such that

$$(2) \qquad \mathscr{L}^n(A) > \delta \mathscr{L}^n(U).$$

Let F be the collection of all closed balls B centered in A and contained in U such that

$$\mathscr{L}^n(A \cap B) < \delta \mathscr{L}^n(B).$$

By (1), F contains arbitrarily small balls centered at each point of A. By the covering theorem, there is a countable disjoint subcollection G covering almost all of A. Therefore,

$$\mathscr{L}^n(A) < \delta \sum_{S \in G} \alpha_n \left(\frac{\operatorname{diam} S}{2} \right)^n \leq \delta \mathscr{L}^n(U).$$

This contradiction of (2) proves the corollary.

2.10. Corollary. *A measurable function $f\colon \mathbf{R}^n \to \mathbf{R}$ is approximately continuous almost everywhere.*

Corollary 2.10 follows rather easily from Corollary 2.9. Exercise 2.9 gives some hints on the proof.

EXERCISES

2.1. Let I be the line segment in $\mathbf{R}^2$ from $(0,0)$ to $(1,0)$. Compute $\mathscr{I}^1(I)$ directly. ($\beta(2,1) = 2/\pi$.)

2.2. Let I be the unit interval $[0,1]$ in $\mathbf{R}^1$. Prove that $\mathscr{H}^1(I) = 1$.

2.3. Prove that $\mathscr{H}^n(\mathbf{B}^n(\mathbf{0}, 1)) < \infty$, just using the definition of Hausdorff measure.

2.4. Let A be a nonempty subset of $\mathbf{R}^n$. First prove that if $0 \le m < k$ and $\mathscr{H}^m(A) < \infty$, then $\mathscr{H}^k(A) = 0$. Second, deduce that the four definitions of the Hausdorff dimension of A are equivalent.

2.5. Define a set $A \subset \mathbf{R}^2$ by starting with an equilateral triangle and removing triangles as follows. Let A_0 be a closed equilateral triangular region of side 1. Let A_1 be the three equilateral triangular regions of side $\frac{1}{3}$ in the corners of A_0. In general let A_{j+1} be the triangular regions, a third the size, in the corners of the triangles of A_j. Let $A = \bigcap A_j$. Prove that $\mathscr{H}^1(A) = 1$.

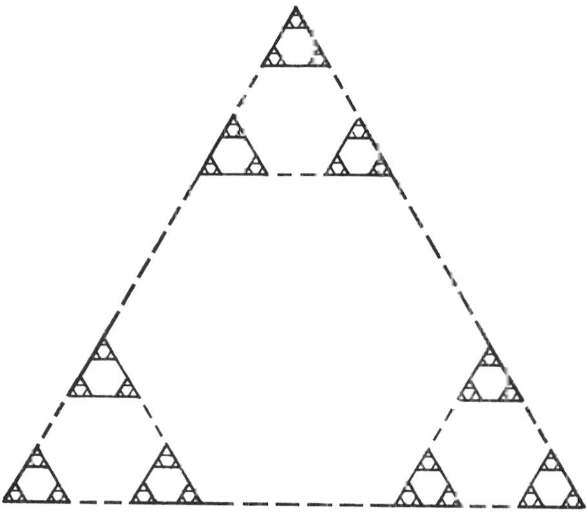

2.6. To define the usual Cantor set $C \subset \mathbf{R}^1$, let $C_1 = [0,1]$; construct C_{j+1} by removing the open middle third of each interval of C_j and put

$$C = \bigcap \{C_j : j \in \mathbb{Z}^+\}.$$

Let $m = \ln 2/\ln 3$.

(a) Prove that $\mathscr{H}^m(C) \le \alpha_m/2^m$ and, hence, $\dim C \le m$.

(b) Try to prove that $\mathscr{H}^m(C) = \alpha_m/2^m$ or at least that $\mathscr{H}^m(C) > 0$ and hence that the Hausdorff dimension of C is m.

2.7. Give a function $f\colon \mathbf{R}^2 \to \mathbf{R}$ which is approximately continuous at $\mathbf{0}$, but for which $\mathbf{0}$ is not a Lebesgue point.

2.8. Prove that if $f\colon \mathbf{R}^m \to \mathbf{R}$ has $\mathbf{0}$ as a Lebesgue point, then f is approximately continuous at $\mathbf{0}$.

2.9. Deduce Corollary 2.10 from Corollary 2.9.
Hint: Let $\{q_i\}$ be a countable dense subset of $\mathbf{R}$, $A_i = \{x : f(x) > q_i\}$, and $E_i = \{x : \Theta(A_i, x) = \chi_{A_i}\}$, and show that f is approximately continuous at each point in $\bigcap E_i$.

CHAPTER 3

Lipschitz Functions and Rectifiable Sets

This chapter introduces the m-dimensional surfaces of geometric measure theory, called rectifiable sets. These sets have folds, corners, and more general singularities. The relevant functions are not smooth functions as in differential geometry, but *Lipschitz* functions.

3.1. Lipschitz Functions. A function $f: \mathbf{R}^m \to \mathbf{R}^n$ is *Lipschitz* if there is a constant C such that

$$|f(x) - f(y)| \le C|x - y|.$$

The least such constant is called the *Lipschitz constant* and is denoted by Lip f. Figure 3.1.1 gives the graphs of two typical Lipschitz functions. Theorems 3.2 and 3.3 show that a Lipschitz function comes very close to being differentiable.

3.2. Rademacher's Theorem [Federer, 3.1.6]. *A Lipschitz function* $f: \mathbf{R}^m \to \mathbf{R}^n$ *is differentiable almost everywhere.*

The *Proof* has five steps:

(1) A monotonic function $f: \mathbf{R} \to \mathbf{R}$ is differentiable almost everywhere.
(2) Every function $f: \mathbf{R} \to \mathbf{R}$ which is locally of bounded variation (and hence every Lipschitz function) is differentiable almost everywhere.
(3) A Lipschitz function $f: \mathbf{R}^m \to \mathbf{R}^n$ has partial derivatives almost everywhere.

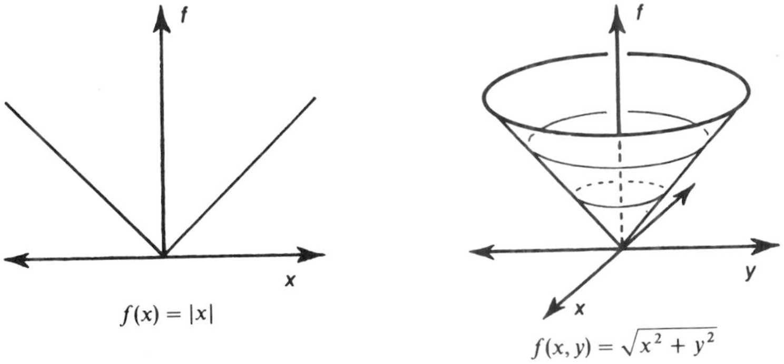

Figure 3.1.1. Examples of Lipschitz functions.

(4) A Lipschitz function f: $\mathbf{R}^m \to \mathbf{R}^n$ is approximately differentiable almost everywhere.
(5) A Lipschitz function f: $\mathbf{R}^m \to \mathbf{R}^n$ is differentiable almost everywhere.

Step (1) is a standard result of real analysis, proved by differentiation of measures. Step (2) follows by decomposing a function of bounded variation as the difference of two monotonic functions. Step (3) follows immediately from Step (2) (modulo checking measurability). The deduction of (4) from (3) is technical, but not surprising, because the existence of continuous partial derivatives implies differentiability and a measurable function is approximately continuous almost everywhere. If (3) holds everywhere, it does not follow that (4) holds everywhere.

The final conclusion (5) rests on the interesting fact that *if a Lipschitz function is approximately differentiable at a, it is differentiable at a*. We conclude this discussion with a proof of that fact.

Suppose that the Lipschitz function f: $\mathbf{R}^m \to \mathbf{R}^n$ is approximately differentiable at a but not differentiable at a. We may assume $a = \mathbf{0}$, $f(\mathbf{0}) = \mathbf{0}$ and $\operatorname{ap} Df(\mathbf{0}) = \mathbf{0}$.

For some $0 < \varepsilon < 1$, there is a sequence of points $a_i \to \mathbf{0}$ such that

$$|f(a_i)| \geq \varepsilon |a_i|.$$

Let $C = \max\{\operatorname{Lip} f, 1\}$. Then, for x in the closed ball $\mathbf{B}(a_i, \varepsilon|a_i|/3C)$,

$$|f(x)| \geq \varepsilon|a_i| - \varepsilon|a_i|/3 \geq \varepsilon|x|/2.$$

Thus, for $x \in E = \bigcup_{i=1}^{\infty} \mathbf{B}(a_i, \varepsilon|a_i|/3C)$,

$$|f(x)| \geq \varepsilon|x|/2.$$

But E does not have density 0 at $\mathbf{0}$, because

$$\frac{\mathscr{L}^m \mathbf{B}(a_i, \varepsilon|a_i|/3C)}{\alpha_m(|a_i| + \varepsilon|a_i|/3C)^m} \geq \frac{(\varepsilon\, a_i|/3C)^m}{(4|a_i|/3)^m} = \frac{\varepsilon^m}{4^m C^m} > 0.$$

This contradiction of the approximate differentiability of f at $\mathbf{0}$ completes the proof.

3.3. Approximation of a Lipschitz Function by a C^1 Function [Federer, 3.1.15]. *Suppose that $A \subset \mathbf{R}^m$ and that $f\colon A \to \mathbf{R}^n$ is Lipschitz. Given $\varepsilon > 0$ there is a C^1 function $g\colon \mathbf{R}^m \to \mathbf{R}^n$ such that $\mathscr{L}^m\{x \in A\colon f(x) \neq g(x)\} \leq \varepsilon$.*

Note that the approximation is in the strongest sense: the functions *coincide* except on a set of measure ε. The proof of 3.3 depends on Whitney's Extension Theorem, which gives the coherence conditions on prescribed values for a desired C^1 function.

3.4. Lemma (Whitney's Extension Theorem) [Federer, 3.1.14]. *Let A be a closed set of points a in $\mathbf{R}^m$ at which the values and derivatives of a desired C^1 function are prescribed by linear polynomials $P_a\colon \mathbf{R}^m \to \mathbf{R}$. For each compact subset C of A and $\delta > 0$, let $\rho(C, \delta)$ be the supremum of the numbers $|P_a(b) - P_b(b)|/|a-b|$, $\|DP_a(b) - DP_b(b)\|$, over all $a, b \in C$ with $0 < |a - b| \leq \delta$. If the prescribed data satisfy the coherence condition that $\lim_{\delta \to 0} \rho(C, \delta) = 0$ for each compact subset C of A, then there exists a C^1 function g satisfying*

$$g(a) = P_a(a),\ Dg(a) = DP_a(a)$$

for all $a \in A$.

Remarks. A more general version of Whitney's Extension Theorem gives the analogous conditions to obtain a C^k function with values and derivatives prescribed by polynomials P_a of degree k. In the proof, the value $g(x)$ assigned at a point $x \notin A$ is a smoothly weighted average of the values prescribed at nearby points of A. The averaging uses a partition of unity subordinate to a cover of A^C which becomes finer and finer as one approaches A.

Sketch of Proof of 3.3. First extend f to a Lipschitz function on all of $\mathbf{R}^m$ (see [Federer [1, 2.10.43]). Second, by Rademacher's Theorem 3.2, f is differentiable almost everywhere. Third, by Lusin's Theorem [Federer, 2.3.5], there is a closed subset E of A such that Df is continuous on E

and $\mathscr{L}^m(A-E)<\varepsilon$. Fourth, for any $a \in E$, $\delta > 0$, define

$$\eta_\delta(a) = \sup_{\substack{0<|x-a|<\delta \\ x\in E}} \frac{|f(x)-f(a)-Df(a)(x-a)|}{|x-a|}$$

Since as $\delta \to 0$, $\eta_\delta \to 0$ pointwise, then by Egoroff's Theorem [Federer, 2.3.7] there is a closed subset F of E such that $\mathscr{L}^m(A-F)<\varepsilon$ and $\eta_\delta \to 0$ uniformly on compact subsets of F. This condition implies the hypotheses of Whitney's Extension Theorem (3.4), with $P_a(x) = f(a) + Df(a)(x-a)$. Consequently there is a C^1 function $g\colon \mathbf{R}^m \to \mathbf{R}^n$ which coincides with f on F.

The following theorem implies for example that Lipschitz images of sets of Hausdorff measure 0 have measure 0.

3.5. Proposition [Federer, 2.10.11]. *Suppose $f\colon \mathbf{R}^l \to \mathbf{R}^n$ is Lipschitz and A is a Borel subset of $\mathbf{R}^l$. Then*

$$\int_{\mathbf{R}^n} N(f|A,y)\,d\mathscr{H}^m y \le (\operatorname{Lip} f)^m \mathscr{H}^m(A).$$

Here $N(f|A,y) \equiv \operatorname{card}\{x \in A\colon f(x)=y\}$.

Proof. Any covering of A by sets S_i of diameter d_i yields a covering of $f(A)$ by the sets $f(S_i)$, of diameter at most $(\operatorname{Lip} f)d_i$. Since the approximating sum $\Sigma\alpha_m(\operatorname{diam}/2)^m$ for the Hausdorff measure contains $(\operatorname{diam})^m$,

$$\mathscr{H}^m(f(A)) \le (\operatorname{Lip} f)^m \mathscr{H}^m(A).$$

Notice that this formula gives the proposition in the case that f is injective. In the general case, chop A up into little pieces A_i and add up the formulas for each piece to obtain

$$\int_{f(A)} (\text{the number of } A_i \text{ intersecting } f^{-1}\{y\})\,d\mathscr{H}^m y \le (\operatorname{Lip} f)^m \mathscr{H}^m(A).$$

As the pieces subdivide, the integrand increases monotonically to the multiplicity function $N(f|A,y)$, and the proposition is proved.

The beginning of this proof illustrates the virtue of allowing coverings by arbitrary sets rather than just balls in the definition of Hausdorff measure. If $\{S_i\}$ covers A, then $\{f(S_i)\}$ is an admissible covering of $f(A)$.

3.6. Jacobians. Jacobians are the corrective factors relating the elements of areas of the domains and images of functions If $f: \mathbf{R}^m \to \mathbf{R}^n$ is differentiable at a, we define the *k-dimensional Jacobian of f at a, $J_k f(a)$*, as the maximum k-dimensional volume of the image under $Df(a)$ of a unit k-dimensional cube.

If rank $Df(a) < k$, $J_k f(a) = 0$. If rank $Df(a) \leq k$, as holds in most applications, then $J_k f(a)^2$ equals the sum of the squares of the determinants of the $k \times k$ submatrices of $Df(a)$. If $k = m$ or n, then $J_k f(a)^2$ equals the determinant of the $k \times k$ product of $Df(a)$ with its transpose. If $k = m = n$, then $J_k f(a)$ is just the absolute value of the determinant of $Df(a)$. In general, computations are sometimes simplified by viewing $Df(a)$ as a map from the orthogonal complement of its kernel onto its image. If $L: \mathbf{R}^m \to \mathbf{R}^m$ is linear, then $\mathscr{L}^m(L(A)) = J_m L \cdot \mathscr{L}^m(A)$.

3.7. The Area Formula [Federer, 3.2.3]. *Consider a Lipschitz function $f: \mathbf{R}^m \to \mathbf{R}^n$ for $m \leq n$.*

(1) *If A is an $\mathscr{L}^m$ measurable set, then*

$$\int_A J_m f(x)\, d\mathscr{L}^m x = \int_{\mathbf{R}^n} N(f|A, y)\, d\mathscr{H}^m y.$$

(2) *If u is an $\mathscr{L}^m$ integrable function, then*

$$\int_{\mathbf{R}^m} u(x) J_m f(x)\, d\mathscr{L}^m x = \int_{\mathbf{R}^n} \sum_{x \in f^{-1}\{y\}} u(x)\, d\mathscr{H}^m y.$$

Remark. If f is a smooth embedding, then (1) equates the classical area of the parameterized surface $f(A)$ with the Hausdorff measure of $f(A)$. Therefore for all smooth surfaces, the Hausdorff measure coincides with the classical area.

Sketch of the Proof of the Area Formula 3.7(1). We will split up A into two cases, according to the rank of Df. In either case, by Rademacher's Theorem 3.2 and 3.5, we may assume that f is differentiable.

Case 1. *Df has rank m.* Let $\{s_i\}$ be a countable dense set of affine maps of $\mathbf{R}^m$ onto m-dimensional planes in $\mathbf{R}^n$. Let E_i be a piece of A such that for each $a \in E_i$ the affine functions $f(a) + Df(a)(x-a)$ and $s_i(x)$ are approximately equal. It follows that

(1) $\det s_i \approx J_m f$ on E_i,
(2) f is injective on E_i, and
(3) the associated map from $s_i(E_i)$ to $f(E_i)$ and its inverse both have Lipschitz constant ≈ 1.

Because f is differentiable, the E_i cover A. Refine $\{E_i\}$ into a countable disjoint covering of A by tiny pieces. On each piece E, by (3) and 3.5,

$$\begin{aligned} \mathscr{H}^m(f(E)) &\approx \mathscr{H}^m(s_i(E)) \\ &= \mathscr{L}^m(s_i(E)) \\ &= \int_E \det s_i \, d\mathscr{L}^m \\ &\approx \int_E J_m f \, d\mathscr{L}^m. \end{aligned}$$

Summing over all the sets E yields

$$\int (\text{number of sets } E \text{ intersecting } f^{-1}\{y\})\, d\mathscr{H}^m y \approx \int_A J_m f \, d\mathscr{L}^m.$$

Taking a limit yields

$$\int N(f|A, y)\, d\mathscr{H}^m y = \int_A J_m f \, d\mathscr{L}^m$$

and completes the proof of Case 1.

We remark that it does not suffice in the proof just to cut A up into tiny pieces without using the s_i. Without the requirement that for $a, b \in E$, $Df(a) \approx Df(b)$, f need not even be injective on E, no matter how small E is.

CASE 2. *DF has rank* $< m$. In this case the left-hand side $\int_A J_m f$ is zero. Define a function

$$\begin{aligned} g : \mathbf{R}^m &\to \mathbf{R}^{n+m} \\ x &\to (f(x), \varepsilon x). \end{aligned}$$

Then $J_m(g) \leq \varepsilon(\operatorname{Lip} f + \varepsilon)^{m-1}$. Now by *Case 1*,

$$\begin{aligned}\mathscr{H}^m(f(A)) &\leq \mathscr{H}^m(g(A)) \\ &= \int_A J_m g \\ &\leq \varepsilon(\operatorname{Lip} f + \varepsilon)^{m-1}\mathscr{L}^m(A).\end{aligned}$$

Therefore the right-hand side also must vanish. Finally we remark that 3.7(2) follows from 3.7(1) by approximating u by simple functions.

The following useful formula relates integrals of a function f over a set A to the areas of the level sets $A \cap f^{-1}\{y\}$ of the function.

3.8. The Coarea Formula [Federer, 3.2.11]. *Consider a Lipschitz function $f\colon \mathbf{R}^m \to \mathbf{R}^n$ with $m > n$. If A is an $\mathscr{L}^m$ measurable set, then*

$$\int_A J_n f(x)\, d\mathscr{L}^m x = \int_{\mathbf{R}^n} \mathscr{H}^{m-n}(A \cap f^{-1}\{y\})\, d\mathscr{L}^n y.$$

Proof.

Case 1. *f is orthogonal projection.* If f is orthogonal projection, then $J_n f = 1$, and the coarea formula is reduced to Fubini's Theorem.

General Case. We treat just the main case $J_n f \neq 0$. By subdividing A as in the proof of the area formula, we may assume that f is linear. Then $f = L \circ P$, where P denotes projection onto the n-dimensional orthogonal complement V of the kernel of f and where L is a a nonsingular linear map from V to $\mathbf{R}^n$. Now

$$\begin{aligned}\int_A J_n f\, d\mathscr{L}^m &= |\det L|\,\mathscr{H}^m(A) \\ &= |\det L| \int_{P(A)} \mathscr{H}^{m-n}(P^{-1}\{y\})\, d\mathscr{L}^n y \\ &= \int_{L \circ P(A)} \mathscr{H}^{m-n}((L \circ P)^{-1}\{y\})\, d\mathscr{L}^n y\end{aligned}$$

as desired.

3.9. Tangent Cones. Suppose that $a \in \mathbf{R}^n$, $E \subset \mathbf{R}^n$, and φ is a measure on $\mathbf{R}^n$. Define a measure $\varphi \llcorner E$, "the restriction of φ to E," by

$$(\varphi \llcorner E)(A) = \varphi(E \cap A).$$

As in 2.5, define m-dimensional densities [Federer, 2.10.19]

$$\Theta^m(\varphi, a) = \lim_{r \to 0} \frac{\varphi(\mathbf{B}(a, r))}{\alpha_m r^m}.$$

$$\Theta^m(E, a) = \Theta^m(\mathscr{H}^m \llcorner E, a)$$

$$= \lim_{r \to 0} \frac{\mathscr{H}^m(E \cap \mathbf{B}(a, r))}{\alpha_m r^m}.$$

Define the *tangent cone* of E at a consisting of the *tangent vectors* of E at a:

$$\mathrm{Tan}(E, a) = \{r \in \mathbf{R}: r \geq 0\} \left[\bigcap_{\varepsilon > 0} \mathrm{Clos}\left\{ \frac{x - a}{|x - a|} : x \in E, 0 < |x - a| < \varepsilon \right\} \right]$$

[Federer, 3.1.21].

Define the (smaller) cone of approximate tangent vectors of E at a:

$$\mathrm{Tan}^m(E, a) = \bigcap \{\mathrm{Tan}(S, a): \Theta^m(E - S, a) = 0\}$$

[Federer, 3.2.16]. See Figure 3.9.1.

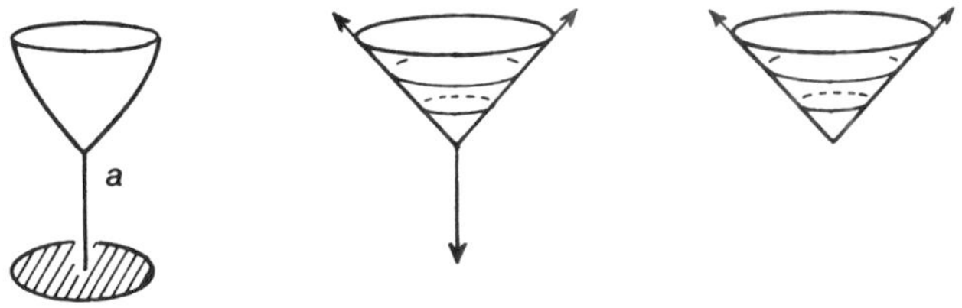

Figure 3.9.1. A set, its tangent cone, and its approximate tangent cone at a. The approximate tangent cone ignores lower-dimensional pieces.

3.10. Rectifiable Sets [Federer, 3.2.14]. A set $E \subset \mathbf{R}^n$ is called $(\mathscr{H}^m, m)$ *rectifiable* if $\mathscr{H}^m(E) < \infty$ and $\mathscr{H}^m$ almost all of E is contained in the union of the images of countably many Lipschitz functions from $\mathbf{R}^m$ to $\mathbf{R}^n$. These sets are the generalized surfaces of geometric measure theory. They include countable unions of immersed manifolds (as long as the total area stays finite) and arbitrary subsets of $\mathbf{R}^m$.

Rectifiable sets can have countably many rectifiable pieces, perhaps connected by countably many tubes and handles and perhaps with all points in $\mathbf{R}^n$ as limit points (cf. Figure 3.10.1). Nevertheless, we will see that from

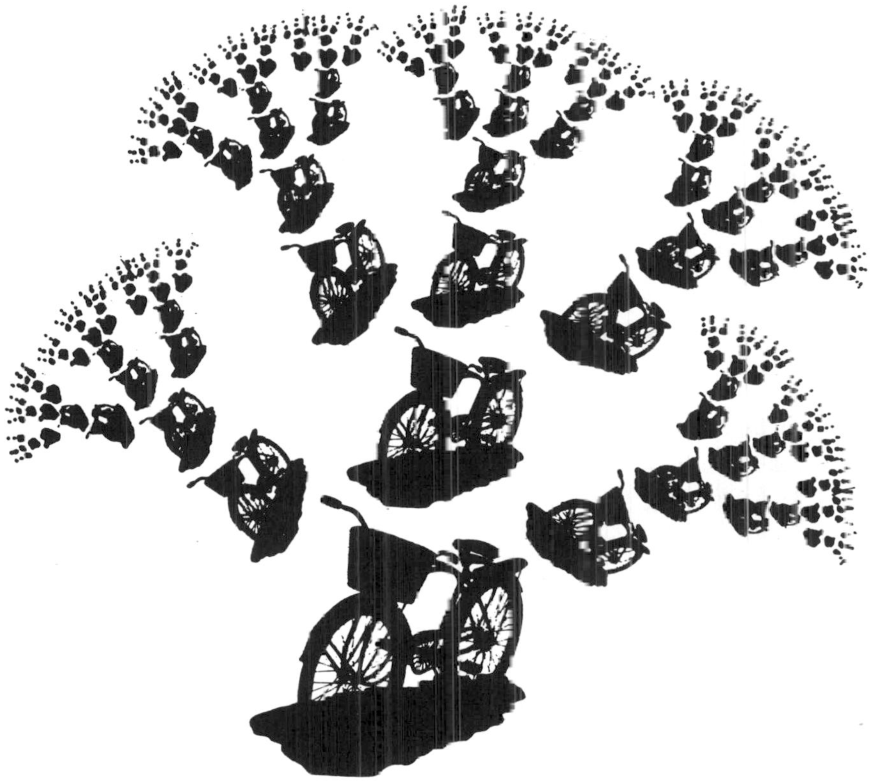

Figure 3.10.1. A two-dimensional rectifiable set in $\mathbf{R}^3$ consisting of the surfaces of countably many bicycles.

the point of view of measure theory, rectifiable sets behave like C^1 submanifolds.

This book will call an $(\mathscr{H}^m, m)$-rectifiable and $\mathscr{H}^m$-measurable set an *m-dimensional rectifiable set*.

The following proposition shows that a measurable set E is rectifiable if and only if $\mathscr{H}^m(E) < \infty$ and $\mathscr{H}^m$ almost all of E is contained in a countable union of C^1, embedded manifolds.

3.11. Proposition [cf. Federer (3.2.18, 3.2.19)]. *In the definition of a rectifiable set E, one can take the Lipschitz functions to be C^1 diffeomorphisms f_j on compact domains with disjoint images whose union coincides with E $\mathscr{H}^m$ almost everywhere. Moreover, the Lipschitz constants of f_j and f_j^{-1} can be taken near* 1.

Proof. It suffices to obtain 1% of the set; the rest can be exhausted by repetition. The first Lipschitz function f can be replaced by a C^1 approximation g by Theorem 3.3. By the area formula 3.7, we may assume Dg is nonsingular. Take just a portion of the domain so that image $g \subset$ image f, Dg is approximately constant, and hence g is injective. Altering domain g by a linear transformation makes $Dg \approx$ identity, and $\text{Lip}\, g \approx \text{Lip}\, g^{-1} \approx 1$. Finally the domain may be replaced by a compact subset. Thus 1% of the set E is obtained. Countably many repetitions prove the proposition.

The following proposition shows that in a certain sense a rectifiable set has a tangent plane at almost every point. (Using different definitions, Hardt and Simon [2, 2.5, p. 22] or Simon [2, 11.6] shows that a modified "rectifiability" is equivalent to the existence of certain "approximate tangent planes" almost everywhere.)

3.12. Proposition [Federer, 3.2.19]. *If W is an m-dimensional rectifiable subset of $\mathbf{R}^n$, then for almost all points a in W, the density $\Theta^m(W, a) = 1$ and* $\text{Tan}^m(W, a)$ *is an m-dimensional plane. If f is a Lipschitz map from W to $\mathbf{R}^\nu$, then f is approximately differentiable $\mathscr{H}^m$ almost everywhere.*

Example. This example gives a modest indication of how bad rectifiable sets can be and hence how strong Proposition 3.12 is. Begin by constructing a Cantor-like set of positive measure as follows. Start with the unit interval. First, remove the middle open interval of length $\frac{1}{4}$. (See Figure 3.12.1.) Second, from the two remaining intervals, remove middle open intervals of total length $\frac{1}{8}$. At the nth step, from the 2^{n-1} remaining intervals, remove middle open intervals of total length $2^{-(n+1)}$. Let C be the intersection. Clearly C contains no interval. However, since the total length removed was $\Sigma 2^{-(n+1)} = \frac{1}{2}$, the length remaining $\mathscr{H}^1(C) = \frac{1}{2}$.

Figure 3.12.1. A Cantor-like set C with $\mathscr{H}^1(C) = \frac{1}{2}$.

Now define $g\colon [0, 1] \to \mathbf{R}^2$ by

$$g(x) = (x, \text{dist}(x, C)).$$

See Figure 3.12.2.

Figure 3.12.2. The image of g intersects $[0, 1]$ in the set C.

Then image g and hence $E = [0, 1] \cup (\text{image } g)$ are rectifiable, even though E fails to be a submanifold at all points of C. Nevertheless, Proposition 3.12 says that $\Theta^1(E, x) = 1$ and $\operatorname{Tan}^1(E, x)$ is a line at almost all points $x \in C$.

Remarks on Proof. The proof that $\Theta^m(W, a) = 1$ almost everywhere uses a covering argument (see Corollary 2.9).

Proposition 3.11 implies that $\operatorname{Tan}^m(W, a)$ contains an m-plane almost everywhere. Since $\Theta^m(W, a) = 1$, it can contain no more.

Similarly by Proposition 3.11, at almost every point, neglecting sets of density 0, W is parameterized by a nonsingular C^1 map $g\colon \mathbf{R}^m \to \mathbf{R}^n$. By Rademacher's Theorem, 3.2, $f \circ g$ is differentiable almost everywhere, and hence f is approximately differentiable almost everywhere.

Here we state a general theorem which subsumes both the area and the coarea formula.

3.13. General Area–Coarea Formula [Federer, 3.2.22]. *Let W be an m-dimensional rectifiable subset of $\mathbf{R}^n$, Z a μ-dimensional rectifiable subset of $\mathbf{R}^\nu$, $m \geq \mu \geq 1$, and f a Lipschitz function from W to Z. Then*

$$\int_W \operatorname{ap} J_\mu f \, d\mathscr{H}^m = \int_Z \mathscr{H}^{m-\mu}(f^{-1}\{z\}) \, d\mathscr{H}^\mu z.$$

More generally, for any $\mathscr{H}^m \llcorner W$ integrable function g on W,

$$\int_W g \cdot \operatorname{ap} J_\mu f \, d\mathscr{H}^m = \int_Z \int_{f^{-1}\{z\}} g \, d\mathscr{H}^{m-\mu} \, d\mathscr{H}^\mu z.$$

3.14. Product of measures [Federer, 3.2.23]. *Let W be an m-dimensional rectifiable Borel subset of $\mathbf{R}^n$ and let Z be a μ-dimensional rectifiable Borel subset of $\mathbf{R}^\nu$. If W is contained in the image of a single Lipschitz function on a bounded subset of $\mathbf{R}^m$, then $W \times Z$ is rectifiable and*

$$\mathscr{H}^{m+\mu} \llcorner (W \times Z) = (\mathscr{H}^m \llcorner W) \times (\mathscr{H}^\mu \llcorner Z).$$

Remarks. In general, the additional hypothesis on W is necessary. If $\mu = \nu$, it holds automatically. In particular, if W is an m-dimensional rectifiable Borel subset of $\mathbf{R}^n$, then $W \times [0,1]^\nu$ is an $(m+\nu)$-dimensional rectifiable subset of $\mathbf{R}^{n+\nu}$. If $m = n$ and $\mu = \nu$, this proposition is just Fubini's Theorem.

The proof, as that of Fubini's Theorem, shows that the collection of sets on which the proposition holds is a σ-algebra.

3.15. Orientation. An *orientation* of an m-dimensional rectifiable subset W of $\mathbf{R}^n$ is a (measurable) choice of orientation for each $\operatorname{Tan}^m(W, a)$. At present no further coherence is required, but we will see in Section 4.2 that a bad choice will make the boundary ∂W much worse. Every rectifiable set of positive measure has uncountably many different orientations (not just two).

3.16. Crofton's Formula [Federer, 3.2.26]. *If W is an m-dimensional rectifiable set, then the integralgeometric measure of W equals its Hausdorff measure:*

$$\mathscr{I}^m(W) = \mathscr{H}^m(W).$$

Remarks. Crofton's Formula follows easily from the coarea formula. The proof, although stated for one-dimensional measure in $\mathbf{R}^2$, applies virtually unchanged to m-dimensional measure in $\mathbf{R}^n$.

Proof. For a one-dimensional measure in $\mathbf{R}^2$,

$$\mathscr{H}^1(W) = \int_W (\text{length of unit tangent})\, d\mathscr{H}^1$$

$$= \int_W \frac{1}{\beta(2,1)} \int_{p \in \mathbf{O}^*(2,1)} (\text{length of projection of unit tangent})\, dp\, d\mathscr{H}^1$$

(because $\mathscr{I}^1$ (unit tangent) $= 1$)

$$= \frac{1}{\beta(2,1)} \int_{p \in \mathbf{O}^*(2,1)} \int_W (\text{length of projection of unit tangent})\, d\mathscr{H}^1\, dp$$

$$= \frac{1}{\beta(2,1)} \int_{p \in \mathbf{O}^*(2,1)} \int_W J_1 p\, d\mathscr{H}^1\, dp$$

$$= \frac{1}{\beta(2,1)} \int_{p \in \mathbf{O}^*(2,1)} \int N(p|W, y)\, d\mathscr{H}^1 y\, dp$$

(by the Coarea Formula, 3.13, because W rectifiable)

$$= \mathscr{I}^1(W).$$

The proof is virtually identical in general dimensions.

3.17. Structure Theorem [Federer, 3.3.13]. This striking theorem describes the structure of arbitrary subsets of $\mathbf{R}^n$. Proved for one-dimensional subsets of $\mathbf{R}^2$ by Besicovitch in 1939, it was generalized to general dimensions by Federer in 1947.

Let E be an arbitrary subset of $\mathbf{R}^n$ with $\mathscr{H}^m(E)<\infty$. Then E can be decomposed as the union of two disjoint sets $E = A \cup B$ with A $(\mathscr{H}^m, m)$ rectifiable and $\mathscr{I}^m(B)=0$.

Remarks. That $\mathscr{I}^m=0$ means that almost all of its projections onto m-planes have measure 0; we might say B is invisible from almost all directions. Such a set B is called *purely unrectifiable*.

The proof, a technical triumph, employs numerous ingenious coverings, notions of density, and amazing dichotomies. A nice presentation of Besicovitch's original proof of the structure theorem for one-dimensional subsets of the plane appears in [Falconer, Chapter 3].

Structure theory had been considered the most daunting component of the proof of the compactness theorem for integral currents, 5.5. In 1986, following Bruce Solomon, Brian White [2] found a direct argument that obviated the dependence on structure theory.

If E is Borel, so are A and B.

Example. Purely unrectifiable sets result from Cantor-type constructions. For example, start with the unit square. Remove a central cross, leaving four squares, each $\frac{1}{4}$ as long as the first. (See Figure 3.17.1.) Similarly, remove central crosses from each small square, leaving 16 smaller squares. Continue, and let the set E be the intersection.

The set E is purely unrectifiable. $\mathscr{H}^1(E)=\sqrt{2}$, but $\mathscr{I}^1(E)=0$. Almost all projections onto lines have measure 0. For example, the projection onto the x-axis is itself a slim Cantor-like set of dimension $\frac{1}{2}$. A diagonal line (with slope $\frac{1}{2}$) gives an exceptional case: the projection is a solid interval. If A is any rectifiable set, then $\mathscr{H}^1(A \cap E)=0$.

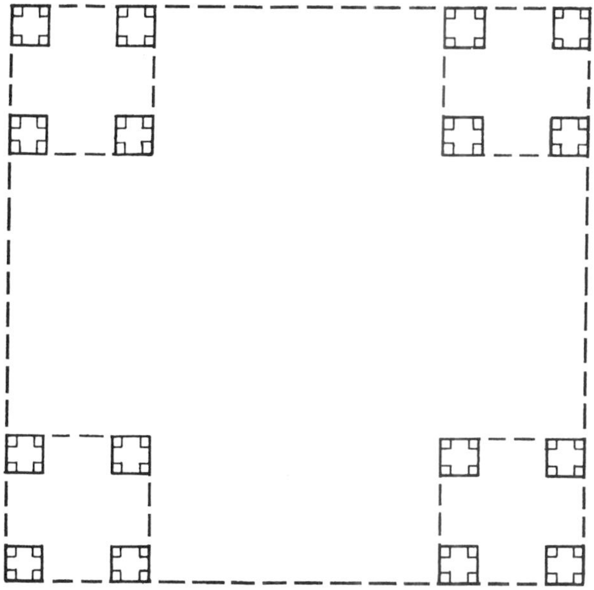

Figure 3.17.1. A purely unrectifiable one-dimensional set E. E is invisible from almost all directions.

EXERCISES

3.1 Give an example of a Lipschitz function $f: [0,1] \to \mathbf{R}$ such that f is not differentiable at any rational point.

3.2. Use Theorem 3.3 to deduce that a Lipschitz function is approximately differentiable almost everywhere.

3.3. Give an example of a continuous function $f: \mathbf{R} \to \mathbf{R}$ such that

(a) given $\varepsilon > 0$ there is a C^1 function $g: \mathbf{R}^1 \to \mathbf{R}^1$ such that $\mathscr{L}^1\{x: f(x) \neq g(x)\} < \varepsilon$, but

(b) f is not Lipschitz.

3.4. Consider the map $f: \mathbf{R}^2 - \{\mathbf{0}\} \to \mathbf{R}^2$ carrying Cartesian coordinates (x, y) to polar coordinates (r, θ). What is $J_1 f$?

3.5. Consider a differentiable map $f: \mathbf{R}^n \to \mathbf{R}$. Show that $J_1 f = |\nabla f|$.

3.6. Compute $\mathscr{H}^2$ of the unit two-sphere $\mathbf{S}^2(0,1)$ by considering the map

$$f: \mathbf{R}^2 \to \mathbf{R}^3$$

$$f: (\varphi, \theta) \to (\sin\varphi\cos\theta, \sin\varphi\sin\theta, \cos\varphi).$$

3.7. Verify the coarea formula for $f: \mathbf{R}^3 \to \mathbf{R}$ given by $f(x, y, z) = x^2 + y^2 + z^2$, $A = \mathbf{B}^3(\mathbf{0}, R)$.

3.8. Let E be an m-dimensional rectifiable Borel subset of the unit sphere in $\mathbf{R}^n$ with $\mathscr{H}^m(E) = a_0$. Let $C = \{tx : x \in E, 0 \le t \le 1\}$.
(a) Rigorously compute $\mathscr{H}^{m+1}(C)$.
(b) Compute $\Theta^{m+1}(C, \mathbf{0})$.
(c) What is $\operatorname{Tan}^{m+1}(C, \mathbf{0})$?

3.9. Give an example of an $(\mathscr{H}^2, 2)$-rectifiable subset E of $\mathbf{R}^3$ which is dense in $\mathbf{R}^3$. Can you also make $\{x \in \mathbf{R}^3 : \Theta^2(E, x) = 1\}$ dense in $\mathbf{R}^3$?

CHAPTER 4

Normal and Rectifiable Currents

In order to define *boundary* and establish compactness properties, it will be useful to view our rectifiable sets as *currents*, i.e., linear functionals on smooth differential forms (named by analogy with electrical currents). The action of an oriented rectifiable set S on a differential form φ is given by integrating the form φ over the set:

$$S(\varphi) = \int_S \varphi \, d\mathscr{H}^m.$$

Currents thus associated with certain rectifiable sets, with integer multiplicities, will be called rectifiable currents. The larger class of normal currents will allow for real multiplicities and smoothing.

The concept of currents is a generalization, by de Rham [2], of distributions. Normal and rectifiable currents are due to Federer and Fleming. Important earlier and contemporaneous work includes the generalized surfaces of L. C. Young [1,2], the frontiers of E. De Giorgi [1,3], and the surfaces of E. R. Reifenberg [1–3]. The general reference for this chapter is [Federer, Chapter IV].

4.1. Vectors and Differential Forms [Federer, Chapter 1 and 4.1]. Consider $\mathbf{R}^n$ with basis $\mathbf{e}_1, \mathbf{e}_2, \ldots, \mathbf{e}_n$. There is a nice way of multiplying m vectors in $\mathbf{R}^n$ to obtain a new object called an *m-vector* ξ:

$$\xi = v_1 \wedge \cdots \wedge v_m.$$

This wedge product is characterized by two properties. First, it is multilinear:

$$cv_1 \wedge v_2 = v_1 \wedge cv_2 = c(v_1 \wedge v_2),$$

$$(u_1 + v_1) \wedge (u_2 + v_2) = u_1 \wedge u_2 + u_1 \wedge v_2 + v_1 \wedge u_2 + v_1 \wedge v_2.$$

Second, it is alternating:

$$u \wedge v = -v \wedge u \quad \text{or} \quad u \wedge u = 0.$$

For example,

$$\begin{aligned}(2\mathbf{e}_1 + 3\mathbf{e}_2 - 5_3) &\wedge (7\mathbf{e}_1 - 11\mathbf{e}_3)\\ &= 14\mathbf{e}_1 \wedge \mathbf{e}_1 - 22\mathbf{e}_1 \wedge \mathbf{e}_3 + 21\mathbf{e}_2 \wedge \mathbf{e}_1 - 33\mathbf{e}_2 \wedge \mathbf{e}_3\\ &\quad - 35\mathbf{e}_3 \wedge \mathbf{e}_1 + 55\mathbf{e}_3 \wedge \mathbf{e}_3\\ &= 0 - 22\mathbf{e}_1 \wedge \mathbf{e}_3 - 21\mathbf{e}_1 \wedge \mathbf{e}_2 - 33\mathbf{e}_2 \wedge \mathbf{e}_3 + 35\mathbf{e}_1 \wedge \mathbf{e}_3 + 0\\ &= -21\mathbf{e}_{12} + 13\mathbf{e}_{13} - 33\mathbf{e}_{23}.\end{aligned}$$

We have abbreviated $\mathbf{e}_{12}$ for $\mathbf{e}_1 \wedge \mathbf{e}_2$.

In general, computation of $\xi = v_1 \wedge \cdots \wedge v_m$ yields an answer of the form

$$\xi = \sum_{i_1 < \cdots < i_m} a_{i_1 \cdots i_m} e_{i_1 \cdots i_m}.$$

The set of all linear combinations of $\{\mathbf{e}_{i_1 \ldots i_m}\colon i_1 < \cdots < i_m\}$ is the space $\Lambda_m \mathbf{R}^n$ of m-vectors, a vectorspace of dimension $\binom{n}{m}$. It has the inner product for which $\{\mathbf{e}_{i_1 \ldots i_m}\}$ is an orthonormal basis.

The purpose of an m-vector $\xi = v_1 \wedge \cdots \wedge v_m$ is to represent the oriented m-plane P through $\mathbf{0}$ of which $v_1, \ldots, v_m$ give an oriented basis. Fortunately, the wedge product $\xi' = v_1' \wedge \cdots \wedge v_m'$ of another oriented basis for P turns out to be a positive multiple of ξ. For example, replacing v_1 with $v_1' = \Sigma c_i v_i$ yields

$$v_1' \wedge v_2 \wedge \cdots \wedge v_m = c_1 v_1 \wedge v_2 \wedge \cdots \wedge v_m.$$

If $v_1, \ldots, v_m$ give an orthonormal basis, then $\xi = v_1 \wedge \cdots \wedge v_m$ has length 1. A product $v_1 \wedge \cdots \wedge v_m$ is 0 if and only if the vectors are linearly dependent. For the case $m = n$,

$$v_1 \wedge \cdots \wedge v_n = \det[v_1, \ldots, v_n] \cdot \mathbf{e}_{1 \cdots n}.$$

An m-vector ξ is called *simple* or *decomposable* if it can be written as a single wedge product of vectors. For example, in $\Lambda_2 \mathbf{R}^4$, $\mathbf{e}_{12} + 2\mathbf{e}_{13} - \mathbf{e}_{23} = (\mathbf{e}_1 + \mathbf{e}_3) \wedge (\mathbf{e}_2 + 2\mathbf{e}_3)$ is simple, whereas $\mathbf{e}_{12} + \mathbf{e}_{34}$ is not (see Exercise 4.5).

The oriented m-planes through the origin in $\mathbf{R}^n$ are in one-to-one correspondence with the unit, simple m-vectors in $\Lambda_m \mathbf{R}^n$.

Incidentally, the geometric relationship between two m-planes in $\mathbf{R}^n$ is given by m angles, as appeared at least as early as [Somerville, IV.12], with beautiful later applications to the geometry of Grassmannians (see Wong) and to area minimization (see Morgan [1, §2.3]).

Let $\mathbf{R}^{n*}$ denote the space of *covectors* dual to $\mathbf{R}^n$, with dual orthonormal basis $\mathbf{e}_1^*, \ldots, \mathbf{e}_n^*$. We remark that dx_j is a common alternate notation for e_j^*. The dual space to $\Lambda_m \mathbf{R}^n$ is the space $\Lambda^m \mathbf{R}^n \equiv \Lambda_m(R^{n*})$ of linear combinations of wedge products of covectors, called m-covectors. The dual basis is $\{e^*_{i_1 \cdots i_m}: i_1 < \cdots < i_m\}$.

A *differential m-form* φ on $\mathbf{R}$ is an m-covectorfield, that is, a map

$$\varphi: \mathbf{R}^n \to \Lambda^m \mathbf{R}^n.$$

For example, one 2-form on $\mathbf{R}^4$ is given by

$$\begin{aligned}\varphi &= \cos x_1 \mathbf{e}_{12}^* + \sin x_1 \mathbf{e}_{34}^* \\ &= \cos x_1 \, dx_1 \, dx_2 + \sin x_1 \, dx_3 \, dx_4 .\end{aligned}$$

The *support*, $\operatorname{spt} \varphi$, of a differential form φ is defined as the closure of $\{x \in \mathbf{R}^n: \varphi(x) \neq 0\}$.

A differential m-form φ is a natural object to integrate over an oriented, m-dimensional rectifiable set S, because it is sensitive to both the location $x \in S$ and the tangent plane to S at x. Let $\vec{S}(x)$ denote the unit m-vector associated with the oriented tangent plane to S at x. Then

$$\int_S \varphi \equiv \int_S \left\langle \vec{S}(x), \varphi(x) \right\rangle d\mathscr{H}^m x.$$

In a classical setting, with no Hausdorff measure available, the definition is more awkward. One uses local parameterizations and proves that the definition is independent of the choice of parameterization. Even the appropriateness of dealing with forms—functions on $\Lambda_m \mathbf{R}^n$— is obscured.

The *exterior derivative* $d\varphi$ of a differential m-form

$$\varphi = \sum f_{i_1 \ldots i_m} \mathbf{e}^*_{i_1 \ldots i_m}$$

is the $(m+1)$-form given by

$$d\varphi = \sum df_{i_1 \ldots i_m} \wedge \mathbf{e}^*_{i_1 \ldots i_m},$$

where $df = (\partial f/\partial x_1)\mathbf{e}_1^* + \cdots + (\partial f/\partial x_n)\mathbf{e}_n^*$. For example, if

$$\varphi = f\,dy\,dz + g\,dz\,dx + h\,dx\,dy,$$

then

$$\begin{aligned} d\varphi &= \left(\frac{\partial f}{\partial x} + \frac{\partial g}{\partial y} + \frac{\partial h}{\partial z}\right) dx\,dy\,dz \\ &= \operatorname{div}(f,g,h)\,dx\,dy\,dz. \end{aligned}$$

If ϕ is a differential l-form and ω is a differential m-form, then

$$d(\phi \wedge \omega) = (d\phi) \wedge \omega + (-1)^m \phi \wedge d\omega.$$

In addition to the dual Euclidean norms $|\xi|$, $|\varphi|$ on $\Lambda_m\mathbf{R}^n$ and $\Lambda^m\mathbf{R}^n$, there are the mass norm $\|\xi\|$ and comass norm $\|\varphi\|^*$, also dual to each other, defined as follows:

$$\|\varphi\|^* = \sup\{|\langle \xi, \varphi\rangle| : \xi \text{ is a unit, simple } m\text{-vector}\};$$

$$\|\xi\| = \sup\{|\langle \xi, \varphi\rangle| : \|\varphi\|^* = 1\}.$$

It follows from convexity theory that

$$\|\xi\| = \inf\left\{\sum |\xi_i| : \xi = \sum \xi_i, \xi_i \text{ simple}\right\}.$$

Consequently, $\|\varphi\|^* = \sup\{|\langle \xi, \varphi\rangle| : \|\xi\| = 1\}$, so that the mass and comass norms are indeed dual to each other. Federer denotes both mass and comass norms by $\|\ \ \|$.

4.2. Currents [Federer, 4.1.1, 4.1.7]. The ambient space is $\mathbf{R}^n$. Let

$$\mathscr{D}^m = \{C^\infty \text{ differential } m\text{-forms with compact support}\}.$$

For example, in $\mathbf{R}^4$, a typical $\varphi \in \mathscr{D}^2$ takes the form

$$\begin{aligned} \varphi &= f_1\,dx_1\,dx_2 + f_2\,dx_1\,dx_3 + f_3\,dx_1\,dx_4 + f_4\,dx_2\,dx_3 + f_5\,dx_2\,dx_4 \\ &\quad + f_6\,dx_3\,dx_4 \\ &= f_1\mathbf{e}_{12}^* + f_2\mathbf{e}_{13}^* + f_3\mathbf{e}_{14}^* + f_4\mathbf{e}_{23}^* + f_5\mathbf{e}_{24}^* + f_6\mathbf{e}_{34}^*, \end{aligned}$$

where the f_j are C^∞ functions of compact support. The topology is generated by locally finite sets of conditions on the f_j and their derivatives of arbitrary order.

The dual space is denoted $\mathscr{D}_m$ and called the space of *m-dimensional currents*. This is a huge space. Under the *weak topology* on $\mathscr{D}_m$, $T_j \to T$ if and only if $T_j(\varphi) \to T(\varphi)$ for all forms $\varphi \in \mathscr{D}^m$.

Any oriented m-dimensional rectifiable set may be viewed as a current as follows. Let $\vec{S}(x)$ denote the unit m-vector associated with the oriented tangent plane to S at x. Then for any differential m-form φ, define

$$S(\varphi) = \int_S \langle \vec{S}(x), \varphi \rangle \, d\mathscr{H}^m$$

Furthermore, we will allow S to carry a positive integer multiplicity $\mu(x)$, with $\int_S \mu(x)\, d\mathscr{H}^m < \infty$, and define

$$S(\varphi) = \int_S \langle \vec{S}(x), \varphi \rangle \mu(x)\, d\mathscr{H}^m.$$

Finally, we will require that S have compact support. Such currents are called *rectifiable currents*.

Definitions for currents are by duality with forms. The boundary of an m-dimensional current $T \in \mathscr{D}_m$ is the $(m-1)$-dimensional current $\partial T \in \mathscr{D}_{m-1}$ defined by

$$\partial T(\varphi) = T(d\varphi).$$

By Stokes's Theorem, this agrees with the usual definition of boundaries if T is (integration over) a smooth oriented manifold with boundary. Notice that giving a piece of the manifold the opposite orientation would create additional boundary. The boundary of a rectifiable current S is generally not a rectifiable current. If it happens to be, then the original current S is called an *integral current*. The *support* of a current is the smallest closed set C such that

$$(\operatorname{spt} \varphi) \cap C = \varnothing \Rightarrow S(\varphi) = 0.$$

4.3. Important Spaces of Currents [Federer, 4.1.24, 4.1.22, 4.1.7, 4.1.5]. The following table gives spaces of currents which play an important role in geometric measure theory.

$\mathscr{P}_m$ integral polyhedral chains	$\subset$	$\mathbf{I}_m$ integral currents	$\subset$	$\mathscr{R}_m$ rectifiable currents	$\subset$	$\mathscr{F}_m$ integral flat chains
$\cap$		$\cap$		$\cap$		$\cap$
$\mathbf{P}_m$ real polyhedral chains	$\subset$	$\mathbf{N}_m$ normal currents	$\subset$	$\mathbf{R}_m$	$\subset$	$\mathbf{F}_m$ real flat chains
						$\cap$
						$\mathscr{E}_m \subset \mathscr{D}_m$

DEFINITIONS. Let

$\mathscr{D}_m = \{m\text{-dimensional currents in } \mathbf{R}^n\}$,

$\mathscr{E}_m = \{T \in \mathscr{D}_m\colon \operatorname{spt} T \text{ is compact}\}$,

$\mathscr{R}_m = \{\text{rectifiable currents}\}$
$= \{T \in \mathscr{E}_m$ associated with oriented rectifiable sets, with integer multiplicities, with finite total measure (counting multiplicites)$\}$,

$\mathscr{P}_m = \{\text{integral polyhedral chains}\}$

= additive subgroup of $\mathscr{E}_m$ generated by classically oriented simplices,

$\mathbf{I}_m = \{\text{integral currents}\}$
$= \{T \in \mathscr{R}_m\colon \partial T \in \mathscr{R}_{m-1}\}$,

$\mathscr{F}_m = \{\text{integral flat chains}\}$
$= \{T + \partial S\colon T \in \mathscr{R}_m,\ S \in \mathscr{R}_{m+1}\}$.

The definitions of the second tier of spaces will appear in Section 4.5.

We also define two important seminorms on the space of currents $\mathscr{D}_m$: the mass $\mathbf{M}$ and the flat norm $\mathscr{F}$.

$$\mathbf{M}(T) = \sup\left\{T(\varphi)\colon \sup_x \|\varphi(x)\|^* \leq 1\right\},$$

$$\mathscr{F}(T) = \inf\left\{\mathbf{M}(A) + \mathbf{M}(B)\colon T = A + \partial B,\ A \in \mathscr{R}_m,\ B \in \mathscr{R}_{m+1}\right\}.$$

The mass of a rectifiable current is just the Hausdorff measure of the associated rectifiable set (counting multiplicities). Note that the norm $\sup_x \|\varphi(x)\|^*$ gives a weaker topology on $\mathscr{D}^m$ than the one to which currents are dual, so that a general current may well have infinite mass. Similarly, $\mathscr{F}(T) < \infty$ if and only if $T \in \mathscr{F}_m$.

The flat norm gives a good indication of when surfaces are geometrically close together. For example, in Figure 4.3.1,

$$\mathbf{M}(S - T) = 2,$$

whereas

$$\mathscr{F}(S - T) \leq \mathbf{M}(A) + \mathbf{M}(B) = 2\varepsilon + \varepsilon = 3\varepsilon.$$

The flat norm topology is clearly weaker than the mass norm topology but stronger than the weak topology. Actually it turns out that for integral cur-

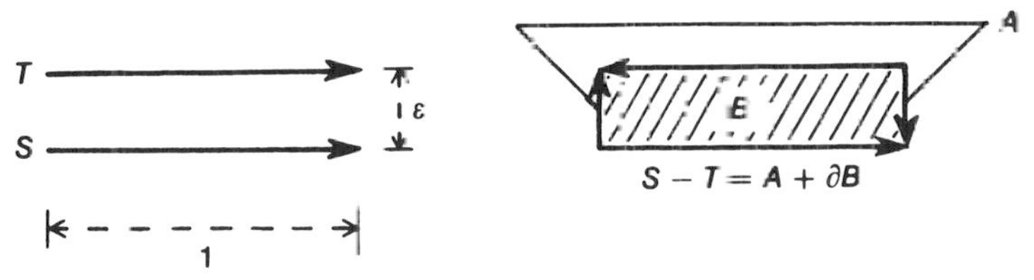

Figure 4.3.1. When two curves or surfaces, S and T, are close together, the flat norm of their difference is small. Here by definition $\mathscr{F}(S-T) \leq \mathbf{M}(A)+\mathbf{M}(B) = 3\varepsilon$.

rents of bounded mass and boundary mass, the flat and weak topologies coincide (cf. Simon [2, 31.2]).

To obtain a rectifiable current which is not an integral current, choose the underlying rectifiable set E with infinite boundary. For example, let E be a connected open subset of the unit disc bounded by a curve of infinite length, as in Figure 4.3.2.

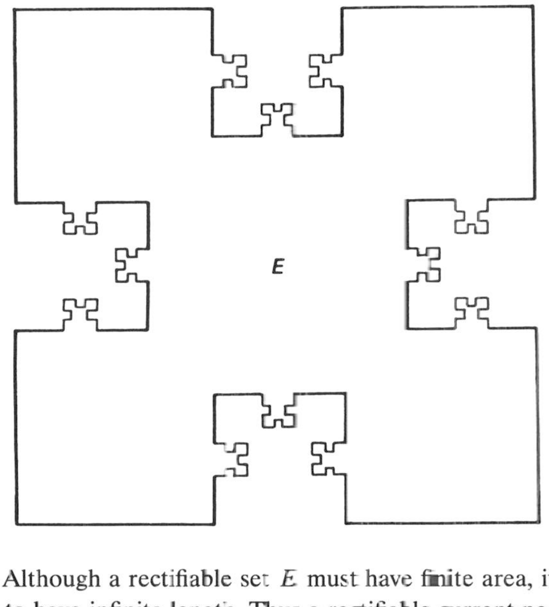

Figure 4.3.2. Although a rectifiable set E must have finite area, its boundary can wiggle enough to have infinite length. Thus a rectifiable current need not be an integral current. Here the length of each successive smaller square rapidly approaches one-third the length of the larger square.

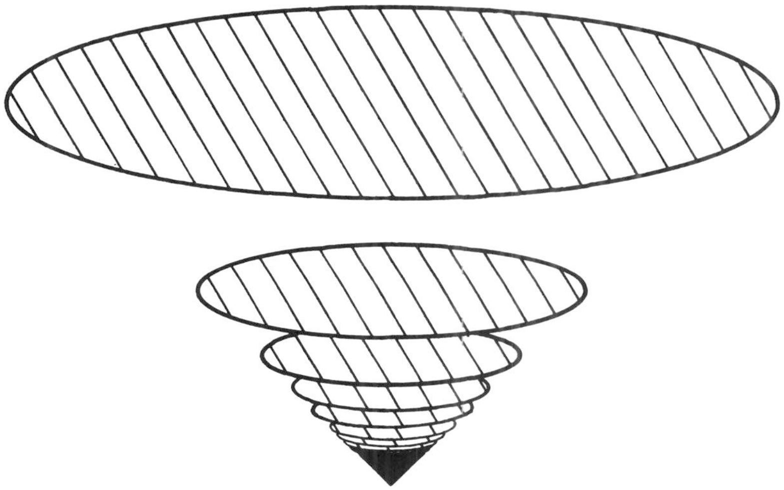

Figure 4.3.3. This infinite collection of discs gives another example of a rectifiable current which is not an integral current. There is finite total area, but infinite total boundary length.

Alternatively, let E be a countable union of discs of radius $1/k$:

$$E = \bigcup_{k \in \mathbf{Z}^+} \{(x, y, z) : x^2 + y^2 \leq k^{-2},\ z = k^{-1}\}.$$

See Figure 4.3.3.

In both cases, the associated rectifiable current T is not an integral current, and ∂T is an integral flat chain but not a rectifiable current.

Actually, only by having infinite boundary mass can a rectifiable current fail to be an integral current. The difficult Closure Theorem, 5.4, will show that

$$(1) \qquad \begin{aligned} \mathbf{I}_m &= \{T \in \mathscr{R}_m : \mathbf{M}(\partial T) < \infty\}, \\ \mathscr{R}_m &= \{T \in \mathscr{F}_m : \mathbf{M}(T) < \infty\}. \end{aligned}$$

(The equivalence of these two equalities follows immediately from the definitions.)

Remarks on Supports and Notation. Let K be a compact C^1 submanifold of $\mathbf{R}^n$, with or without boundary (or more generally, a "compact Lipschitz neighborhood retract"). Federer uses the subscript K to denote support in K. For example,

$$\mathscr{R}_{m,K} = \{T \in \mathscr{R}_m : \operatorname{spt} T \subset K\}.$$

(For arbitrary compact K, $\mathscr{R}_{m,K}$ has a more technical meaning [Federer, 4.1.29].)

Similarly, a norm $\mathscr{F}_K$ is defined by

$$\mathscr{F}_K(T) \equiv \inf\{\mathbf{M}(A) + \mathbf{M}(B) : T = A + \partial B,\ A \in \mathscr{R}_{m,K},\ B \in \mathscr{R}_{m+1,K}\}.$$

If K is any large ball containing $\operatorname{spt} T$, then $\mathscr{F}_K(T)$ equals what we have called $\mathscr{F}(T)$, as can be seen by projecting the A and B from the definition of $\mathscr{F}$ onto K. In the other main case of interest, when K is a compact C^1 submanifold of $\mathscr{R}^n$, $\mathscr{F}_K(T) \geq \mathscr{F}(T)$, with strict inequality sometimes. However, $\mathscr{F}_K$ and $\mathscr{F}$ yield the same topology on $\mathscr{F}_{m,K}$, the integral flat chains supported in K.

Mapping Currents. Next we want to define the image of a compactly supported current under a C^∞ map $f: \mathbf{R}^n \to \mathbf{R}^\nu$. First, for any simple m-vector $\xi = v_1 \wedge \cdots \wedge v_m \in \Lambda_m \mathbf{R}^n$, and point x in the domain of f, define the push-forward of ξ in $\Lambda_m \mathbf{R}^\nu$:

$$[\Lambda_m(Df(x))](\xi) = (Df(x))(v_1) \wedge \cdots \wedge (Df(x))(v_m).$$

The map $\Lambda_m(Df(x))$ extends to a linear map on all m-vectors.

Second, for any differential m-form $\varphi \in \mathscr{D}^m(\mathbf{R}^\nu)$, define its pullback $f^\# \varphi$ on $\mathbf{R}^n$ by

$$\langle \xi, f^\# \varphi(x) \rangle = \langle [\Lambda_m(Df(x))](\xi), \varphi(f(x)) \rangle.$$

Finally, for any compactly supported current $T \in \mathscr{D}_m(\mathbf{R}^n)$, define its push-forward $f_\# T \in \mathscr{D}_m(\mathbf{R}^\nu)$ by

$$(f_\# T)(\varphi) = T(f^\# \varphi).$$

If T is the rectifiable current associated with some oriented rectifiable set E, then $f_\# T$ is the rectifiable current associated with the oriented rectifiable set $f(E)$, with the appropriate multiplicities (see Exercise 4.23). The boundary $\partial(f_\# T) = f_\# \partial T$. In many cases the smoothness hypothesis on f may be relaxed.

A current $T \in \mathscr{D}_m$ is called *representable by integration* if there is a Borel regular measure $\|T\|$ on $\mathbf{R}^n$, finite on compact sets, and a function $\vec{T}: \mathbf{R}^n \to \Lambda_m \mathbf{R}^n$ with $\|\vec{T}(x)\| = 1$ for $\|T\|$ almost all x such that

$$T(\varphi) = \int \langle \vec{T}(x), \varphi(x) \rangle \, d\|T\| x.$$

We write $T = \|T\| \wedge \vec{T}$. A current $T \in \mathscr{D}_m$ of finite mass is automatically representable by integration, as follows from the representation theory of general measure theory. On the other hand, the current $T \in \mathscr{D}_m(\mathbf{R}^n)$ defined by

$$T(a_1 dx_1 \wedge \cdots \wedge dx_m + \cdots) = \frac{\partial a_1}{\partial x_1}(p),$$

where p is a fixed point in $\mathbf{R}^n$, has infinite mass, and is not representable by integration.

Every rectifiable current S is representable by integration. Indeed, if E is the associated set with multiplicity function l, then $\|S\|$ is the measure $l(\mathscr{H}^m \llcorner E)$ and $\vec{S}$ is the unit m-vectorfield orienting E. $S = l(\mathscr{H}^m \llcorner E) \wedge \vec{S} = (\mathscr{H}^m \llcorner E) \wedge \eta$, where $\eta = l\vec{S}$.

4.4. Theorem [Federer, 4.1.28]. *The following are equivalent definitions for $T \in \mathscr{E}_m$ to be a rectifiable current.*

(1) *Given $\varepsilon > 0$, there are an integral polyhedral chain $P \in \mathscr{P}_m(\mathbf{R}^\nu)$ and a Lipschitz function $f: \mathbf{R}^\nu \to \mathbf{R}^n$ such that*

$$\mathbf{M}(T - f_\# P) < \varepsilon.$$

(2) *There are a rectifiable set B and an $\mathscr{H}^m \llcorner B$ summable m-vectorfield η such that η is simple, $|\eta(x)|$ is an integer ("the multiplicity"), $\operatorname{Tan}^m(B, x)$ is associated with $\eta(x)$, and $T(\varphi) = \int_B \langle \eta(x), \varphi(x) \rangle \, d\mathscr{H}^m$.*

Remarks. In (1), if T is supported in a closed ball K, one may assume $\operatorname{spt} f_\# P \subset K$, by replacing $f_\# P$ by its projection onto K. Actually, Federer takes (1) as the definition of $\mathscr{R}_m$, whereas we have used (2).

A current $(\mathscr{H}^m \llcorner B) \wedge \eta$ can fail to be rectifiable in several ways: the set B could fail to be rectifiable or to have compact closure; the total mass $\int_B \|\eta(x)\| \, d\mathscr{H}^m$ could fail to be finite; the given m-vector $\eta(x)$ could fail to be tangent to B at x; or $|\eta(x)|$ could fail to be an integer.

Proof Sketch. First support (1) holds. Since each side of the polyhedral chain is a subset of some $\mathbf{R}^m$, its image under f is rectifiable and hence $f_\# P$ is a rectifiable current and satisfies (2). But now T, as a mass convergent sum of such, obtained by successive approximation, is a rectifiable current.

The opposite implication depends on the following lemma of measure theory.

LEMMA. *Let A be a bounded ($\mathscr{L}^m$-measurable) subset of $\mathbf{R}^m$. Then given $\varepsilon > 0$, there is a finite disjoint set of m-simplices which coincide with A except for a set of measure less than ε.*

Proof of Lemma. We may assume that A is open, by replacing A by a slightly larger open set. Cover 1% of A by disjoint simplices (as in the proof of the Besicovitch Covering Theorem, 2.7). Repeat on what is left. After N repetitions, $1-(.99)^N$ of A is covered by disjoint simplices, as desired.

Completion of Proof of Theorem. Suppose T satisfies (2). The rectifiable set B is the union of Lipschitz images of subsets of $\mathbf{R}^m$. Use the lemma to approximate B by images of polyhedra.

4.5. Normal Currents [Federer, 4.1.7, 4.1.12]. In preparation for the definitions of more general spaces of currents, we define a more general flat norm, $\mathbf{F}$. For any current $T \in \mathscr{D}_m$, define

$$\begin{aligned}\mathbf{F}(T) &= \sup\{T(\varphi): \varphi \in \mathscr{D}^m,\ \|\varphi(x)\|^* \leq 1 \quad \text{and} \quad \|d\varphi(x)\|^* \leq 1 \text{ for all } x\} \\ &= \min\{\mathbf{M}(A) + \mathbf{M}(B): T = A + \partial B,\ A \in \mathscr{E}_m,\ B \in \mathscr{E}_{m+1}\}.\end{aligned}$$

The second equality shows the similarity of the norm $\mathbf{F}$ and the previously defined norm $\mathscr{F}$. Inequality ($\leq$) is easy, since if $T = A + \partial B$ as in the minimum and φ is as in the supremum, then

$$T(\varphi) = (A + \partial B)(\varphi) = A(\varphi) + B(d\varphi) \leq \mathbf{M}(A) + \mathbf{M}(B).$$

Equality is proved using the Hahn–Banach Theorem (Federer [1, p. 367]).

Now continuing the definitions of the spaces of currents in the diagram in the beginning of Section 4.3, let

$$\mathbf{N}_m = \{T \in \mathscr{E}_m: \mathbf{M}(T) + \mathbf{M}(\partial T) < \infty\}$$

$$= \{T \in \mathscr{E}_m: T \text{ and } \partial T \text{ are representable by integration}\},$$

$$\mathbf{F}_m = \mathbf{F}\text{-closure of } \mathbf{N}_m \text{ in } \mathscr{E}_m,$$

$$\mathbf{R}_m = \{T \in \mathbf{F}_m: \mathbf{M}(T) < \infty\},$$

$$\mathbf{P}_m = \{\text{real linear combinations of elements of } \mathscr{P}_m\}.$$

The important space $\mathbf{N}_m$ of normal currents allows real densities and smoothing. For example, if A is the unit square region

$$\{(x, y): 0 \leq x \leq 1, 0 \leq y \leq 1\}$$

in the plane, then $S_1 = \sqrt{2}(\mathscr{H}^2 \llcorner A) \wedge \mathbf{e}_{12}$ is a two-dimensional normal current which is not an integral current (See Figure 4.5.1.) S_1 is $\sqrt{2}$ times the integral current $(\mathscr{H}^2 \llcorner A) \wedge \mathbf{e}_{12}$. $S_2 = (\mathscr{H}^2 \llcorner A) \wedge \mathbf{e}_1$ is a one-dimensional

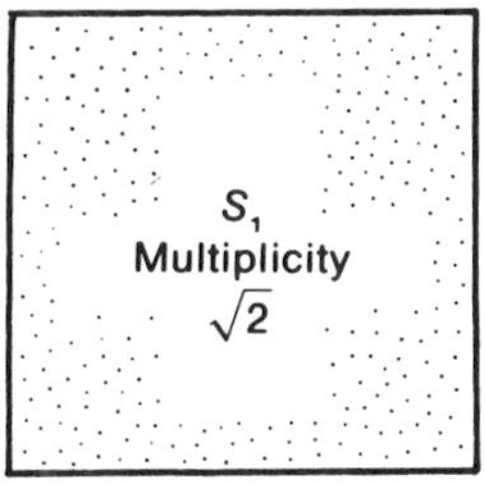

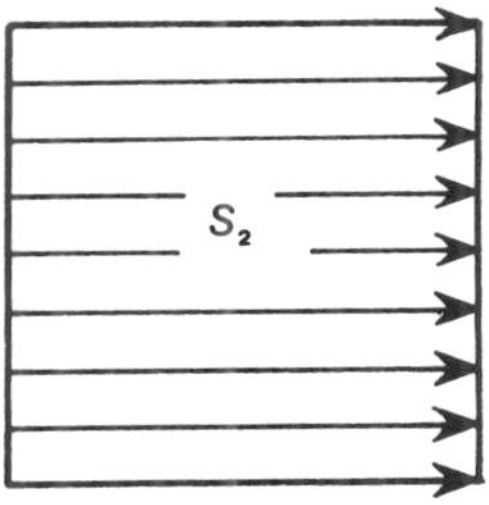

Figure 4.5.1. Currents with noninteger densities and one-dimensional currents spread over two-dimensional sets gives examples of normal currents which are not integral currents.

normal current which is not an integral current. To check that S_2 is indeed a normal current, compute ∂S_2 from the definition:

$$\begin{aligned}\partial S_2(f(x,y)) = S_2(df) &= S_2\left(\frac{\partial f}{\partial x}\mathbf{e}_1^* + \frac{\partial f}{\partial y}\mathbf{e}_2^*\right)\\ &= \int_A \left\langle \mathbf{e}_1, \frac{\partial f}{\partial x}\mathbf{e}_1^* + \frac{\partial f}{\partial y}\mathbf{e}_2^* d\mathscr{H}^2\right.\\ &= \int_A \frac{\partial f}{\partial x}\, dx\, dy\\ &= \int_0^1 [f(1,y) - f(0,y)]\, dy\\ &= \int_0^1 f(1,y)\, dy - \int_0^1 f(0,y)\, dy.\end{aligned}$$

Therefore,

$$\partial S_2 = \mathscr{H}^1 \llcorner \{(1,y): 0 \leq y \leq 1\} - \mathscr{H}^1 \llcorner \{(0,y): 0 \leq y \leq 1\},$$

and $\mathbf{M}(\partial S_2) = 2 < \infty$. If $B = \{(x,0): 0 \leq x \leq 1\}$, $T = \mathscr{H}^1 \llcorner B \wedge \mathbf{e}_1$, and $\tau_{(x,y)}$ denotes translation by (x,y), then

$$S_2 = \int_0^1 \tau_{(0,y)\#} T\, dy.$$

Thus S_2 is an integral of integral currents.

More generally, if T is any m-dimensional integral current in $\mathbf{R}^n$ and f is a function of compact support with $\int |f|\, d\mathscr{L}^n < \infty$, then the weighted smoothing of T

$$S = \int_{x \in \mathbf{R}^n} f(x) \cdot \tau_{x\#} T \, d\mathscr{L}^n x$$

is a normal current. Of course,

$$\partial S = \int_{x \in \mathbf{R}^n} f(x) \cdot \tau_{x\#} \partial T \, d\mathscr{L}^n x.$$

Whether every normal current can be written as an integral of integral currents has been a subject of research. A counterexample was provided by M. Zworski.

4.6. Proposition [Federer, 4.1.17]. *The space* $\mathbf{R}_m$ *is the* $\mathbf{M}$*-closure of* $\mathbf{N}_m$ *in* $\mathscr{E}_m$.

Proof. Clearly $\mathbf{R}_m$ is $\mathbf{M}$-closed in $\mathscr{E}_m$. Suppose $T \in \mathbf{R}_m$. Given $\varepsilon > 0$, choose $S \in \mathbf{N}_m$ such that $\mathbf{F}(T - S) < \varepsilon$. Hence there are currents $A \in \mathscr{E}_m$ and $B \in \mathscr{E}_{m+1}$ such that $T - S = A + \partial B$ and $\mathbf{M}(A) + \mathbf{M}(B) < \varepsilon$. Since $\mathbf{M}(\partial B) = \mathbf{M}(T - S - A) < \infty$, $\partial B \in \mathbf{N}_m$. Therefore $S + \partial B \in \mathbf{N}_m$, and $\mathbf{M}(T - (S + \partial B)) = \mathbf{M}(A) < \varepsilon$. Hence T is the $\mathbf{M}$-closure of $\mathbf{N}_m$, as desired.

We have seen examples of m-dimensional normal currents based on higher-dimensional sets. The following theorem shows that even real flat chains cannot be supported in lower-dimensional sets. The hypothesis that the integral geometric measure $\mathscr{I}^m(\operatorname{spt} T) = 0$ holds if the Hausdorff measure $\mathscr{H}^m(\operatorname{spt} T) = 0$, as follows easily from the definition of $\mathscr{I}^m$(2.4).

4.7. Theorem [Federer, 4.1.20]. *If* $T \in \mathbf{F}_m(\mathbf{R}^n)$ *and* $\mathscr{I}^m(\operatorname{spt} T) = 0$, *then* $T = 0$.

Examples. The current $S \equiv \mathscr{H}^0 \llcorner \{(0,0)\} \wedge \mathbf{e}_1 \in \mathscr{D}_1$ is not flat because $\mathscr{I}^1(\operatorname{spt} S) = \mathscr{I}^1\{(0,0)\} = 0$. The current

$$T = \mathscr{H}^1 \llcorner \{(0,y): 0 \leq y \leq 1\} \wedge \mathbf{e}_1$$

is not flat, because if it were, its projection on the x-axis, which is S, would be flat. (See Figure 4.7.1.) This example illustrates the principle that for a flat current, the prescribed vectorfield must lie down "flat" (see Federer [1, 4.1.15]). The suggestiveness of the term *flat* is a happy accident. H. Whitney, also a student of music, coined the term for the smaller of his flat and sharp norms, originally designated $\| \; \|_\flat$, and $\| \; \|_\sharp$.

Figure 4.7.1. The current T is not flat; its prescribed vectorfield is not tangent to the underlying set.

Outline of Proof

I. *Smoothing.* A smooth normal current in $\mathbf{R}^n$ is one of the form $\mathscr{L}^n \wedge \xi$, with ξ a smooth m-vectorfield of compact support. Any normal current T can be approximated in the flat norm by a smooth normal current $T_\varepsilon = \mathscr{L}^n \wedge \xi$ as follows. Let f be a smooth approximation to the delta function at $\mathbf{0}$, and put $T_\varepsilon = \int_{x \in \mathbf{R}^n} f(x) \cdot \tau_{x\#} T \, d\mathscr{L}^n x$.

II. *If $T \in \mathbf{F}_n(\mathbf{R}^n)$, then T is of the form $\mathscr{L}^n \wedge \xi$ for some vectorfield ξ* [Federer, 4.1.18]. Notice the assumption of codimension 0, where the norms $\mathbf{F}$ and $\mathbf{M}$ coincide. Therefore T can be $\mathbf{M}$-approximated by a normal current and hence by smoothing by $\mathscr{L}^n \wedge \xi_1$, with ξ_1 a smooth n-vectorfield, $\mathbf{M}(T - \mathscr{L}^n \wedge \xi_1) < 2^{-1}$, and hence

$$\mathbf{M}(\mathscr{L}^n \wedge \xi_1) = \int |\xi_1| \, d\mathscr{L}^n < \mathbf{M}(T) + 2^{-1}.$$

Likewise, $T - \mathscr{L}^n \wedge \xi_1$ can be $\mathbf{M}$-approximated by $\mathscr{L}^n \wedge \xi_2$, with $\mathbf{M}(T - \mathscr{L}^n \wedge \xi_1 - \mathscr{L}^n \wedge \xi_2) < 2^{-2}$, and hence

$$\mathbf{M}(\mathscr{L}^n \wedge \xi_2) = \int |\xi_2| \, d\mathscr{L}^n < 2^{-1} + 2^{-2}.$$

Likewise, $T - \mathscr{L}^n \wedge \xi_1 - \mathscr{L}^n \wedge \xi_2$ can be $\mathbf{M}$-approximated by $\mathscr{L}^n \wedge \xi_3$, with $\mathbf{M}(T - \mathscr{L}^n \wedge \xi_1 - \mathscr{L}^n \wedge \xi_2 - \mathscr{L}^n \wedge \xi_3) < 2^{-3}$, and hence $\mathbf{M}(\mathscr{L}^n \wedge \xi_3) < 2^{-2} + 2^{-3}$. Continue. Since $\int \Sigma_{j=1}^\infty |\xi_j| < \mathbf{M}(T) + 2^{-1} + 2^{-1} + 2^{-2} + 2^{-2} + \cdots = \mathbf{M}(T) + 2 < \infty$, $\Sigma \xi_j$ converges in L^1. Let $\xi = \Sigma \xi_j$. Then $T = \mathscr{L}^n \wedge \xi$ as desired.

III. *Completion of Proof*. For the case $m = n$, the theorem follows immediately from part II. Let $m < n$. Since $\mathscr{I}^m(\operatorname{spt} T) = 0$, we may assume spt T projects to sets of measure 0 in the m-dimensional coordinate axis planes. For notational convenience we take $m = 1$, so that $T \in \mathbf{F}_1(\mathbf{R}^n)$. We consider the action of T on an arbitrary smooth 1-form

$$\varphi = f_1 \mathbf{e}_1^* + f_2 \mathbf{e}_2^* + \cdots + f_n \mathbf{e}_n^*.$$

Since $T(\varphi) = \Sigma T(f_j \mathbf{e}_j^*)$, it suffices to show that $T(f_j \mathbf{e}_j^*) = 0$. Let p_j denote projection onto the jth coordinate axis, and let $T \llcorner f$ denote the current defined by $(T \llcorner f)(\varphi) = T(f\varphi)$ (see 4.11). Then

$$T(f_j \mathbf{e}_j^*) = (T \llcorner f_j)(\mathbf{e}_j^*) = (T \llcorner f_j)\left(p_j^{\#} \mathbf{e}_j^*\right) = \left(p_{j\#}(T \llcorner f_j)\right)(\mathbf{e}_j^*).$$

Since $p_{j\#}(T \llcorner f_j) \in \mathbf{F}_m(\mathbf{R}^m)$ is of the form $\mathscr{L}^m \wedge \xi$ by part II, and its support has measure 0, it must be 0. Therefore $T(f_j \mathbf{e}_j^*) = 0$, as desired.

4.8. Theorem [Federer, 4.1.23]. *Given a real flat chain $T \in \mathbf{F}_m$ and $\varepsilon > 0$, there is a real polyhedral approximation $P \in \mathbf{P}_m$ satisfying $\mathbf{F}(T - P) \leq \varepsilon$ and $\mathbf{M}(P) \leq \mathbf{M}(T) + \varepsilon$.*

Proof. Since the space $\mathbf{F}_m$ is defined as the $\mathbf{F}$-closure of $\mathbf{N}_m$, and if $\mathbf{M}(T) < \infty$, T lies in the $\mathbf{M}$-closure of $\mathbf{N}_m$ (Proposition 4.6), we may assume $T \in \mathbf{N}_m$. By smoothing (cf. proof of 4.7, part I), we may assume T is of the form $T = \mathscr{L}^n \wedge \xi(x)$, where $\xi(x)$ is a smooth m-vectorfield of compact support with $\int |\xi(x)|\, d\mathscr{L}^n < \infty$. By approximating ξ by step functions, we may assume T is of the form $T = \mathscr{L}^n \llcorner A \wedge \eta$, for some bounded set A and m-vector η. We may assume $\eta = \mathbf{e}_{1 \cdots m}$ and A is the unit cube $\{0 \leq x_i \leq 1\} \subset \mathbf{R}^n$. Now we can approximate $T = \mathscr{L}^n \llcorner A \wedge \eta$ by layers. Take a large integer, M, let

$$B = \{x \in \mathbf{R}^m : 0 \leq x_i \leq 1\} \times \left\{\frac{1}{M}, \frac{2}{M}, \ldots, 1\right\}^{n-m} \subset \mathbf{R}^n,$$

and let

$$P = M^{-(n-m)}(\mathscr{L}^m \llcorner B) \wedge \mathbf{e}_{1 \cdots m}.$$

Then $\mathbf{M}(P) = \mathbf{M}(T)$ and for M large, $\mathbf{F}(T - P) < \varepsilon$.

4.9. Constancy Theorem [Federer, 4.1.31]. *Suppose B is an m-dimensional connected, C^1 submanifold with boundary of $\mathbf{R}^n$, classically oriented* *by ζ. If a real flat chain $T \in \mathbf{F}_m$ is supported in B and its boundary is supported in the boundary of B, then, for some real number r,*

$$T = r(\mathscr{H}^m \llcorner B) \wedge \zeta.$$

Of course if T is an integral flat chain, then r is an integer.

Proof. We must show locally that $\partial T = 0$ means T is constant. We may assume locally that $B = \mathbf{R}^m \times \{\mathbf{0}\} \subset \mathbf{R}^n$. Then T is of the form $\mathscr{L}^m \wedge \xi$ for some m-vectorfield $\xi = f \cdot \mathbf{e}_{1 \cdots m}$ (proof of 4.7, part II). For any smooth $(m-1)$-form

$$\varphi = g_1 \mathbf{e}^*_{2 \cdots m} - g_2 \mathbf{e}^*_{13 \cdots m} + \cdots g_m \mathbf{e}^*_{12 \cdots m-1}$$

of compact support,

$$\begin{aligned} 0 = \partial T(\varphi) = T(d\varphi) &= \int \langle \xi, d\varphi \rangle \, d\mathscr{L}^m \\ &= \int f \left(\frac{\partial g_1}{\partial x_1} + \frac{\partial g_2}{\partial x_2} + \cdots + \frac{\partial g_m}{\partial x_m} \right) d\mathscr{L}^m \\ &= \int f \operatorname{div} g \, d\mathscr{L}^m. \end{aligned}$$

It follows that f is constant, as desired. (If f is smooth, integration by parts yields that

$$0 = -\int \left(\frac{\partial f}{\partial x_1} g_1 + \cdots + \frac{\partial f}{\partial x_m} g_m \right) d\mathscr{L}^m$$

for all g_j, so that $\partial f / \partial x_i = 0$ and f is constant. For general $f \in L^1$, $\int f \operatorname{div} g = 0$ for all g means the weak derivative vanishes and f is constant.)

4.10. Cartesian Products. Given $S \in \mathscr{D}_m(\mathbf{R}^n)$ and $T \in \mathscr{D}_\mu(\mathbf{R}^\nu)$, one can define their Cartesian product $S \times T \in \mathscr{D}_{m+\mu}(\mathbf{R}^{n+\nu})$. The details appear in Federer [1, 4.1.8, p. 360], but for now it is enough to know that it exists and has the expected properties.

4.11. Slicing [Federer, 4.2.1]. In this section we define $(m-1)$-dimensional slices of m-dimensional normal currents by hyperplanes or by hypersurfaces $\{u(x) = r\}$. It will turn out that for almost all values of r, the slices themselves are normal currents.

First, for any current $T \in \mathscr{D}_m$ and C^∞ differential k-form α, define a current $T \llcorner \alpha \in \mathscr{D}_{m-k}$ by

$$(T \llcorner \alpha)(\varphi) = T(\alpha \wedge \varphi).$$

In particular, if α is a function (0-form) f, then $(T \llcorner f)(\varphi) = T(f\varphi)$. The symbol $\llcorner$ for such "interior multiplication," sometimes called "elbow," points to the term of lower degree which gets pushed to the other side in the definition.

If T is representable by integration, $T = \|T\| \wedge \vec{T}$, then it suffices to assume that $\int |f| d\|T\| < \infty$. Indeed, then $T \llcorner f = f\|T\| \wedge \vec{T}$: one just multiplies the multiplicity by f. Of course, even if T is rectifiable, $T \llcorner f$ will not be, unless f is integer valued. For $A \subset \mathbf{R}^n$, define "T restricted to A," $T \llcorner A = T \llcorner \chi_A$, where χ_A is the characteristic function of A.

For a normal current $T \in \mathbf{N}_m\mathbf{R}^n$, a Lipschitz function u: $\mathbf{R}^n \to \mathbf{R}$, and a real number r, define the slice

(1) $$\langle T, u, r+\rangle \equiv (\partial T) \llcorner \{x: u(x) > r\} - \partial(T \llcorner \{x: u(x) > r\})$$
$$= \partial(T \llcorner \{x: u(x) \le r\}) - (\partial T) \llcorner \{x: u(x) \le r\}.$$

(See Figure 4.11.1.) It follows that

(2) $$\partial\langle T, u, r+\rangle = -\langle \partial T, u, r+\rangle.$$

Proposition

(3) $$\mathbf{M}\langle T, u, r+\rangle \le (\operatorname{Lip} u) \varliminf_{h \to 0+} \|T\|\{r < u(x) < r+h\}/h.$$

In particular, if $f(r) = \|T\| B(x,r)$, then for almost all r,

$$\mathbf{M}\langle T, u, r+\rangle \le f'(r).$$

Proof. If χ is the characteristic function of the set $\{x: u(x) > r\}$, then

$$\langle T, u, r+\rangle = (\partial T) \llcorner \chi - \partial(T \llcorner \chi).$$

For small, positive h, approximate χ by a C^∞ function f satisfying

$$f(x) = \begin{cases} 0 & \text{if } u(x) \le r \\ 1 & \text{if } u(x) \ge r+h \end{cases}$$

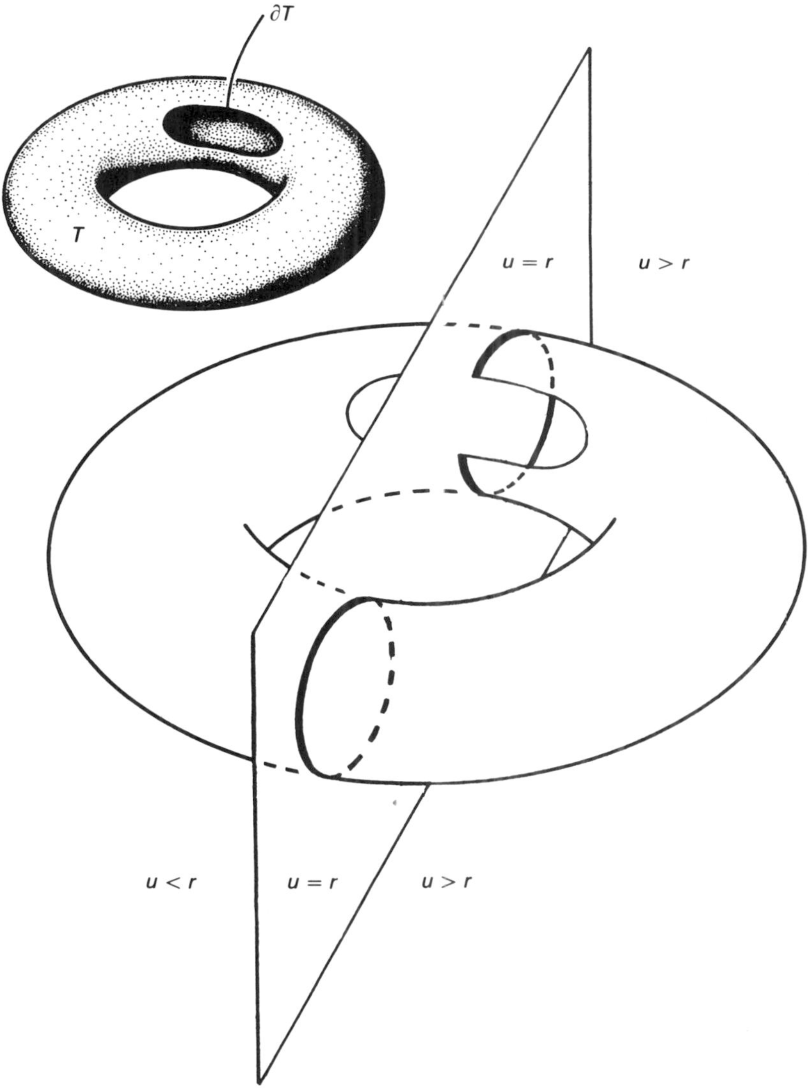

Figure 4.11.1. The slice of the torus T by the pictured plane consists of $1\frac{3}{4}$ circles.

and $\operatorname{Lip} f \lesssim (\operatorname{Lip} u)/h$. Then

$$\begin{aligned}\mathbf{M}\langle T, u, r+\rangle &\approx \mathbf{M}((\partial T)\llcorner f - \partial(T\llcorner f)) \\ &= \mathbf{M}(T\llcorner df) \\ &\leq (\operatorname{Lip} f)\|T\|\{x\colon r < u(x) < r+h\} \\ &\lesssim (\operatorname{Lip} u)\|T\|\{x\colon r < u(x) < r+h\}/h.\end{aligned}$$

Consequently,

$$\mathbf{M}\langle T,u,r+\rangle \le (\operatorname{Lip} u)\ \underline{\lim}_{h\to 0} \|T\|\{x: r<u(x)<r+h\}/h.$$

PROPOSITION

$$(4)\qquad \int_a^b \mathbf{M}\langle T,u,r+\rangle\, d\mathscr{L}^1 r \le (\operatorname{Lip} u)\|T\|\{x: a<u(x)<b\}.$$

Proof. Consider the function $f(r)=\|T\|\{u(x)<r\}$. Since f is monotonically increasing, its derivative, $f'(r)$, exists for almost all r. Hence

$$\begin{aligned}(\operatorname{Lip} u)\|T\|\{a<u(x)<b\} &= (\operatorname{Lip} u)\Big(f(b)-\lim_{x\to a+} f(x)\Big)\\ &\ge (\operatorname{Lip} u)\int_a^b f'(r)\,dr \ge \int_a^b \mathbf{M}\langle T,u,r+\rangle\,dr\end{aligned}$$

by (3).

COROLLARY

$$(5)\qquad \langle T,u,r+\rangle \in \mathbf{N}_{m-1}$$

for almost all r.

The corollary follows directly from (4) and (2) (see Exercise 4.20). Of course, it follows that if T is rectifiable, so are almost all slices.

PROPOSITION

$$(6)\qquad \int_a^b \mathbf{F}[T\llcorner\{u(x)\le r\}]\, d\mathscr{L}^1 r \le [b-a+\operatorname{Lip} u]\mathbf{F}(T).$$

For a hint on the proof, see Exercise 4.21.

Remarks. If T is an integral current, so is almost every slice, as will follow from the Closure Theorem, 5.4(2, 3), and 4.11(5) (or as is shown directly in Simon [2, §28]). Slicing can be generalized to a vector-valued function $u\colon \mathbf{R}^n \to \mathbf{R}^l$ [Federer, 4.3].

The following lemma considers slices of T by the function $u(x)=|x-a|$. If T has no boundary, then

$$\langle T,u,r+\rangle \equiv \partial(T\llcorner\{x: u(x)\le r\}) = \partial(T\llcorner\mathbf{B}(a,r)).$$

The lemma says that if almost all such "slices by spheres" are rectifiable, then T is rectifiable.

4.12. Lemma [Federer, 4.2.15]. If T is a normal current without boundary and if, for each $a \in \mathbf{R}^n$, $\partial(T \llcorner \mathbf{B}(a,r))$ is rectifiable for almost all $r \in \mathbf{R}$, then T is rectifiable.

Remarks. This lemma is crucial to proving the Closure and Compactness Theorems of Chapter 5. The proof of this lemma in Federer uses structure theory and a covering argument. In 1986, following the work of Solomon, White [2] discovered a way of circumvent the structure theory at almost no cost.

EXERCISES

4.1. Compute $(\mathbf{e}_1 + 2\mathbf{e}_2 + 3\mathbf{e}_3) \wedge (\mathbf{e}_1 + 2\mathbf{e}_2 - 3\mathbf{e}_3) \wedge \mathbf{e}_4$.

4.2. Consider the 2-plane P in $\mathbf{R}^4$ given by

$$P = \{(x_1, x_2, x_3, x_4): x_1 + x_2 + x_3 = x_3 + x_4 = 0\}.$$

Find a nonorthogonal basis u, v and an orthonormal basis w, z for P. Verify by direct computation that $u \wedge v$ is a multiple of $w \wedge z$ and that $|w \wedge z| = 1$.

4.3. Verify by direct computation that

$$(\mathbf{e}_1 + 2\mathbf{e}_2 + 3\mathbf{e}_3) \wedge (\mathbf{e}_1 - \mathbf{e}_3) \wedge (\mathbf{e}_2 + \mathbf{e}_3) = \begin{vmatrix} 1 & 1 & 0 \\ 2 & 0 & 1 \\ 3 & -1 & 1 \end{vmatrix} \mathbf{e}_{123}.$$

4.4. Prove that the 2-vector $\mathbf{e}_{12} + 2\mathbf{e}_{13} + 2\mathbf{e}_{23}$ is simple.

4.5. Prove that $\mathbf{e}_{12} + \mathbf{e}_{34}$ is not simple.

4.6. Find the integral of the differential form

$$\varphi = x_1(\sin x_1 x_2)\mathbf{e}_{12}^* + e^{x_1 + x_2 + x_3}\mathbf{e}_{13}^* + \mathbf{e}_{23}^*$$

over $\{(x_1, x_2, x_3) \in \mathbf{R}^3: 0 \le x_1 \le 1, 0 \le x_2 \le 1, x_3 = 0\}$ (with the usual upward orientation).

4.7. Find the integral of the differential form

$$\varphi = 2\mathbf{e}_{12}^* + 3\mathbf{e}_{13}^* + 5\mathbf{e}_{23}^*$$

over $\{(x_1, x_2, x_3) \in \mathbf{R}^3: x_1 + x_2 + x_3 = 0 \text{ and } x_1^2 + x_2^2 + x_3^2 \le 1\}$.

4.8. Prove that the boundary operator ∂ maps $\mathbf{I}_m$ into $\mathbf{I}_{m-1}$ and $\mathscr{F}_m$ into $\mathscr{F}_{m-1}$. Also prove that $\operatorname{spt} \partial T \subset \operatorname{spt} T$.

4.9. For the rectifiable currents $T \in \mathscr{R}_1(\mathbf{R}^2)$ and for C^∞ functions f, g, and h, compute formulas for $T(f\,dx + g\,dy)$ and $\partial T(h)$.

(*a*) $T = \mathscr{H}^1 \llcorner \{(x,0): 0 \le x \le 1\} \wedge \mathbf{e}_1$

(*b*) $T = (\mathscr{H}^1 \llcorner \{(x,x): 0 \le x \le 1\}) \wedge 3\sqrt{2}\,(\mathbf{e}_1 + \mathbf{e}_2)$.

4.10. Prove that $\mathbf{I}_m$ is $\mathbf{M}$ dense in $\mathscr{R}_m$ and $\mathscr{F}$ dense in $\mathscr{F}_m$.

4.11. Prove that $\{T \in \mathscr{R}_m\colon \operatorname{spt} T \subset \mathbf{B}(0,R)\}$ is $\mathbf{M}$ complete and that $\{T \in \mathscr{F}_m\colon \operatorname{spt} T \subset \mathbf{B}(0,R)\}$ is $\mathscr{F}$ complete.

4.12. Prove that ∂ carries $\mathbf{N}_m$ into $\mathbf{N}_{m-1}$ and $\mathbf{F}_m$ into $\mathbf{F}_{m-1}$.

4.13. Check this analogy to 4.3(1):

$$\mathbf{N}_m = \{T \in \mathbf{R}_m\colon \mathbf{M}(\partial T) < \infty\},$$

$$\mathbf{R}^m = \{T \in \mathbf{F}_m\colon \mathbf{M}(T) < \infty\}.$$

4.14. Prove that, in analogy with the definitions of $\mathbf{I}_m$ and $\mathscr{F}_m$,

$$\mathbf{N}_m = \{T \in \mathbf{R}_m\colon \partial T \in \mathbf{R}_{m-1}\}$$

$$\mathbf{F}_m = \{T + \partial S\colon T \in \mathbf{R}_m,\ S \in \mathbf{R}_{m+1}\}.$$

4.15. Prove that $\mathbf{I}_m \subset \mathbf{N}_m$, $\mathscr{R}_m \subset \mathbf{R}_m$, and $\mathscr{F}_m \subset \mathbf{F}_m$.

4.16. For the currents $T \in \mathscr{D}_1(\mathbf{R}^2)$ representable by integration and for C^∞ functions f, g, and h,

(i) write a formula for $T(f\,dx + g\,dy)$,

(ii) write a formula for $\partial T(h)$, and

(iii) give the smallest space of currents from the table at the beginning of Section 4.3 to which T belongs.

(a) $T = \sum_{k=1}^{\infty} \mathscr{H}^1 \llcorner \{(k^{-1}, y)\colon 0 \le y \le 2^{-k}\} \wedge \mathbf{j}$.

(b) $T = \mathscr{H}^2 \llcorner \{(x,y)\colon 0 \le x \le 1,\ 0 \le y \le 1\} \wedge \mathbf{i}$.

(c) $T = \mathscr{H}^1 \llcorner \{(x,0)\colon 0 \le x \le 1\} \wedge \mathbf{j}$.

(d) $T = \mathscr{H}^0 \llcorner \{a\} \wedge \mathbf{i}$.

(e) $T = \mathscr{H}^2 \llcorner \{(x,y)\colon x^2 + y^2 \le 1\} \wedge \mathbf{i}$.

4.17. Let E be the modification of the Cantor set obtained by starting with the unit interval and removing 2^{n-1} middle intervals, each of length 4^{-n}, at the nth step $(n = 1, 2, 3 \ldots)$.

(a) Show that $\mathscr{H}^1(E) = \frac{1}{2}$.

(b) Show that $\mathscr{H}^1 \llcorner E \wedge \mathbf{i}$ is a rectifiable current, but not an integral current.

4.18. Prove the second equality in 4.11(1) above.

4.19. Prove 4.11(2).

4.20. Deduce 4.11(5) from 4.11(4) and 4.11(2).

4.21. Prove 4.11(6).

Hint: First show that, if $T = A + \partial B$ with T, A, and $B \in \mathbf{N}$, then

$$T \llcorner \{u(x) \le r\} = A \llcorner \{u(x) \le r\} + \partial\left[B \llcorner \{u(x) \le r\}\right] - \langle B, u, r- \rangle.$$

4.22. Prove that $\mathbf{M}$ is $\mathbf{F}$ lower semicontinuous on $\mathscr{D}_m$, i.e., if $T_i, T \in \mathscr{D}_m$, and $T_i \xrightarrow{\mathbf{F}} T$, then $\mathbf{M}(T) \le \liminf \mathbf{M}(T_i)$.

Hint: Work right from the definition of $\mathbf{M}$ in 4.3 and the first definition of $\mathbf{F}$ in 4.5.

4.23. Suppose f is a C^∞ map from $\mathbf{R}^n$ to $\mathbf{R}^\nu$ and $S = l(\mathscr{H}^m \llcorner E) \wedge \vec{S}$ is a rectifiable current represented by integration in terms of an underlying rectifiable set E and an integer-valued multiplicity function l.

(a) Assuming that f is injective, show that

$$f_\# S = l \circ f^{-1}(\mathscr{H}^m \llcorner f(E)) \wedge \frac{(\Lambda_m Df)(\vec{S})}{|(\Lambda_m Df)(\vec{S})|}.$$

(b) Without assuming that f is injective, show that

$$f_\# S = (\mathscr{H}^m \llcorner f(E)) \wedge \sum_{y = f(x)} l(x) \frac{(\Lambda_m Df(x))(\vec{S})}{|(\Lambda_m Df(x))(\vec{S})|}.$$

Hint: Use the definitions and the general coarea formula, 3.13.

CHAPTER 5

The Compactness Theorem and the Existence of Area-Minimizing Surfaces

The Compactness Theorem, 5.5, deserves to be known as the fundamental theorem of geometric measure theory. It guarantees solutions to a wide class of variational problems in general dimensions. It says that a certain set $\mathcal{T}$ of surfaces is compact in a natural topology. The two main lemmas are the Deformation Theorem, 5.1, which will imply in 5.2 that $\mathcal{T}$ is totally bounded, and the Closure Theorem, 5.4, which will imply that $\mathcal{T}$ is complete.

5.1. The Deformation Theorem [Federer, 4.2.9]. The Deformation Theorem approximates an integral current T by deforming it onto a grid of mesh $2\varepsilon > 0$. (See Figure 5.1.1.) The resulting approximation, P, is automatically a polyhedral chain. The main error term is ∂S, where S is the surface through which T is deformed. There is a secondary error term, Q, due to moving ∂T into the skeleton of the grid.

Whenever $T \in \mathbf{I}_m\mathbf{R}^n$ *and* $\varepsilon > 0$, *there exist* $P \in \mathscr{P}_m\mathbf{R}^n$, $Q \in \mathbf{I}_m\mathbf{R}^n$, *and* $S \in \mathbf{I}_{m+1}\mathbf{R}^n$ *such that the following conditions hold with* $\gamma = 2n^{2m+2}$:

(1) $$T = P + Q + \partial S.$$

(2) $$\begin{aligned} \mathbf{M}(P) &\le \gamma[\mathbf{M}(T) + \varepsilon\mathbf{M}(\partial T)], \\ \mathbf{M}(\partial P) &\le \gamma\mathbf{M}(\partial T), \\ \mathbf{M}(Q) &\le \varepsilon\gamma\mathbf{M}(\partial T), \\ \mathbf{M}(S) &\le \varepsilon\gamma\mathbf{M}(T). \end{aligned}$$

Consequently, $\mathscr{F}(T - P) \le \varepsilon\gamma(\mathbf{M}(T) + \mathbf{M}(\partial T))$.

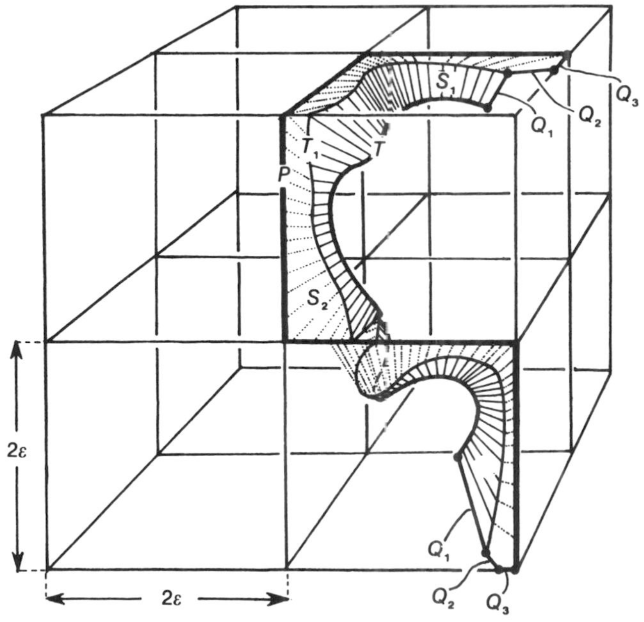

Figure 5.1.1. The Deformation Theorem describes a multistep process for deforming a given curve T onto a polygon P in the 2ε grid. During the process surfaces S_1 and S_2 are swept out. The endpoints of T trace out curves Q_1, Q_2, and Q_3.

(3) spt P *is contained in the m-dimensional* 2ε *grid; i.e., if* $x \in \operatorname{spt} P$, *then at least* $n-m$ *of its coordinates are even multiples of* ε. *Also,* $\operatorname{spt} \partial P$ *is contained in the* $(m-1)$*-dimensional* 2ε *grid.*

(4) $$\operatorname{spt} P \cup \operatorname{spt} Q \cup \operatorname{spt} S \subset \{x : \operatorname{dist}(x, \operatorname{spt} T) \le 2n\varepsilon\}.$$

Proof Sketch, Case $m = 1$, $n = 3$. Let W_k denote the k-dimensional ε grid:

$$W_k = \{(x_1, x_2, x_3) \in \mathbf{R}^3 : \text{at least } 3-k \text{ of the } x_j \text{ are even multiples of } \varepsilon\}.$$

Then W_2 consists of the boundaries of $2\varepsilon \times 2\varepsilon \times 2\varepsilon$ cubes.

First project the curve T radially outward from the centers of the cubes onto W_2. (For now, suppose T stays away from the centers.) Let S_1 be the surface swept out by T during this projection, let Q_1 be the curve swept out by ∂T, and let T_1 be the image of T in W_2. Then, with suitable orientations,

$$T = T_1 + Q_1 + \partial S_1.$$

The mass of T_1 is of the same order as the mass of T, $\mathbf{M}(T_1) \sim \mathbf{M}(T)$. Likewise $\mathbf{M}(\partial T_1) \sim \mathbf{M}(\partial T)$, $\mathbf{M}(Q_1) \sim \varepsilon \mathbf{M}(\partial T)$, and $\mathbf{M}(S_1) \sim \varepsilon \mathbf{M}(T)$.

Second, W_1, the 1-skeleton of W_2, consists of the boundaries of $2\varepsilon \times 2\varepsilon$ square regions. Project the curve T_1 radially outward from the centers of the squares onto W_1. Let S_2 be the surface swept out by T_1, let Q_2 be the curve swept out by ∂T_1, and let T_2 be the image of T_1 in W_1. Then, $T_1 = T_2 + Q_2 + \partial S_2$, and the masses are of order $\mathbf{M}(T_2) \sim \mathbf{M}(T_1) \sim \mathbf{M}(T)$, $\mathbf{M}(\partial T_2) \sim \mathbf{M}(\partial T)$, $\mathbf{M}(Q_2) \sim \varepsilon \mathbf{M}(\partial T)$, and $\mathbf{M}(S_2) \sim \varepsilon \mathbf{M}(T)$.

Third, let Q_3 consist of line segments from each point of ∂T_2 to the nearest point in the 0-skeleton W_0. Put $P = T_2 - Q_3$. Then not only does P lie in W_1, but also ∂P lies in W_0. In particular, P is an integral polyhedral chain. The masses satisfy

$$\mathbf{M}(Q_3) \sim \varepsilon \mathbf{M}(\partial T),$$

$$\mathbf{M}(P) = \mathbf{M}(T_2) + \mathbf{M}(Q_3) \sim \mathbf{M}(T) + \varepsilon \mathbf{M}(\partial T),$$

$$\mathbf{M}(\partial P) = \mathbf{M}(\partial T_2) \sim \mathbf{M}(\partial T).$$

Let $Q = Q_1 + Q_2 + Q_3$ and $S = S_1 + S_2$. Then $T = P + Q + \partial S$. The masses satisfy

$$\mathbf{M}(Q) = \mathbf{M}(Q_1) + \mathbf{M}(Q_2) + \mathbf{M}(Q_3) \sim \varepsilon \mathbf{M}(\partial T),$$

$$\mathbf{M}(S) \sim \varepsilon \mathbf{M}(T),$$

completing the proof.

There is one problem with the foregoing sketch. If the original curve winds tightly about (or, worse, passes through) one of the centers of radial projection, the mass of its projection could be an order larger than its own mass. In this case, one moves the curve a bit before starting the whole process. If the original curve winds throughout space, it may be impossible to move it away from the centers of projection, but the average distortion can still be controlled.

5.2. Corollary. *The set*

$$\mathscr{T} = \{T \in \mathbf{I}_m : \operatorname{spt} T \subset \mathbf{B}^n(\mathbf{0}, c_1),\ \mathbf{M}(T) \le c_2,\ \text{and } \mathbf{M}(\partial T) \le c_3\}$$

is totally bounded under $\mathscr{F}$.

Proof. Each $T \in \mathscr{T}$ can be well approximated by a polyhedral chain P in the ε-grid with $\mathbf{M}(P) \leq \gamma[c_2 + \varepsilon c_3]$ and $\operatorname{spt} P \subset \mathbf{B}^n(\mathbf{0}, c_1 + 2n\varepsilon)$ by 5.1(4). Since there are only finitely many such P, $\mathscr{T}$ is totally bounded.

5.3. The Isoperimetric Inequality [Federer, 4.2.10]. *If $T \in \mathbf{I}_m\mathbf{R}^n$ with $\partial T = 0$, then there exists $S \in \mathbf{I}_{m+1}\mathbf{R}^n$ with $\partial S = T$ and*

$$\mathbf{M}(S)^{m/(m+1)} \leq \gamma \mathbf{M}(T).$$

Here $\gamma = 2n^{2m+2}$ as in the Deformation Theorem.

Remarks. That T bounds some rectifiable current S is shown by taking the cone over T. The value of the isoperimetric inequality lies in the numerical estimate on $\mathbf{M}(S)$. It was long conjectured that the worst case (exhibiting the best constant) was the sphere, in all dimensions and codimensions. This conjecture was proven in 1986 by Almgren [2]. For merely stationary surfaces, an isoperimetric inequality still holds [Allard, §7.1], but the sharp constant remains conjectural, even for minimal surfaces in $\mathbf{R}^3$. For more general (bounded) integrands than area, an isoperimetric in-

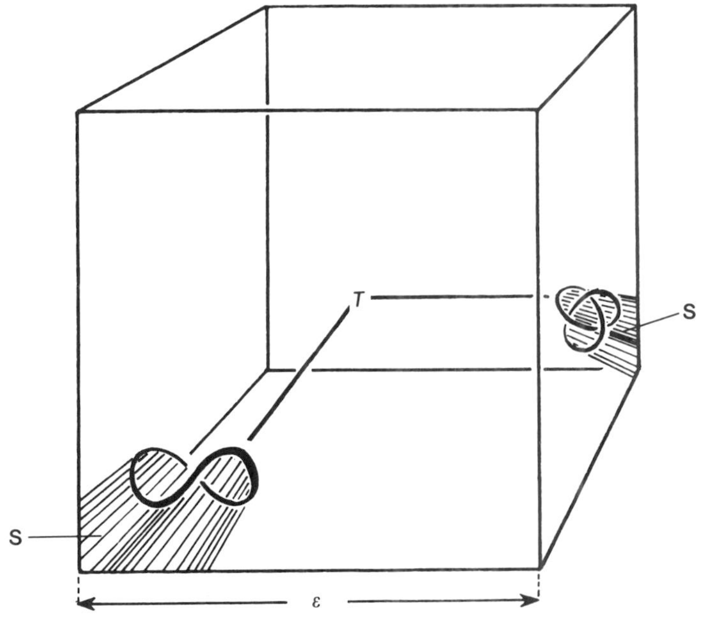

Figure 5.3.1. In the proof of the Isoperimetric Inequality, T is projected onto a *large* grid.

equality follows trivially for minimizers, but remains conjectural for stationary surfaces, even in $\mathbf{R}^3$.

The proof of 5.3 is a bizarre application of the Deformation Theorem, $T = P + Q + \partial S$. One chooses ε *large*, a grid large enough to force the approximation P to T to be 0. See Figure 5.3.1.

Proof. Choose ε so that $\gamma \mathbf{M}(T) = \varepsilon^m$, and apply the Deformation Theorem to obtain $T = P + Q + \partial S$. Since $\partial T = 0$, $Q = 0$. By 5.1(2), $\mathbf{M}(P) \leq \gamma \mathbf{M}(T)$. But, since P lies in the 2ε grid, $\mathbf{M}(P)$ must be a multiple of $(2\varepsilon)^m$, which exceeds $\gamma \mathbf{M}(T)$ by choice of ε. Therefore $P = 0$. Now $T = \partial S$, and, by 5.1(2),

$$\mathbf{M}(S) \leq \varepsilon \gamma \mathbf{M}(T) = \varepsilon^{m+1} = [\gamma \mathbf{M}(T)]^{(m+1)/m}.$$

5.4. The Closure Theorem [Federer, 4.2.16].

(1) $\mathbf{I}_m$ *is* $\mathbf{F}$ *closed in* $\mathbf{N}_m$,
(2) $\mathbf{I}_{m+1} = \{T \in \mathscr{R}_{m+1} : \mathbf{M}(\partial T) < \infty\}$,
(3) $\mathscr{R}_m = \{T \in \mathscr{F}_m : \mathbf{M}(T) < \infty\}$.
Consequently,
(4) $\mathscr{T} = \{T \in \mathbf{I}_m : \operatorname{spt} T \subset \mathbf{B}^n(\mathbf{0}, R),\ \mathbf{M}(T) \leq c,$ *and* $\mathbf{M}(\partial T) \leq c\}$, *is* $\mathscr{F}$ *complete*.

Remarks. This result, specifically (1), is the hard part of the Compactness Theorem. It depends on Lemma 4.12, the characterization of rectifiable sets by rectifiable slices. The original proof of Lemma 4.12 used structure theory. However, in 1986 White [2] found a simpler and more direct argument.

It follows directly from the definition of $\mathscr{F}_m$ that $\{T \in \mathscr{F}_m : \operatorname{spt} T \subset \mathbf{B}(\mathbf{0}, R)\}$ is $\mathscr{F}$ complete. Consequence (4) now follows by the lower semicontinuity of $\mathbf{M}$ (Exercise 4.22) and the continuity of ∂.

Proof. We leave it as an exercise (5.2) to check that for each m, (1) $\Rightarrow$ (2) $\Rightarrow$ (3). Hence to prove (1), (2), and (3), it suffices to prove (1) for m, assuming all three conclusions for $m-1$.

To prove (1), suppose that a sequence of integral currents $Q_i \in \mathbf{I}_m$ converges in the $\mathbf{F}$ norm to a normal current $T \in \mathbf{N}_m$. We must show that $T \in \mathbf{I}_m$.

By induction, we may assume that $\partial T \in \mathbf{I}_{m-1}$. By replacing T with $T - T_1$, where $T_1 \in \mathbf{I}_m$ has the same boundary as T, we may assume that $\partial T = 0$.

By Lemma 4.12, it suffices to show that for all points $p \in \mathbf{R}^n$, for almost every positive real number r, the slice $\partial(T \llcorner \mathbf{B}(p,r))$ is rectifiable. We may assume that the Q_i's converge so rapidly that

$$\Sigma \mathbf{F}(Q_i - T) < \infty.$$

Thence by slicing theory 4.11(6), for $0 < a < b$,

$$\int_a^b \sum \mathbf{F}[(Q_i - T) \llcorner \mathbf{B}(p,r)]\, dr < \infty.$$

Therefore $Q_i \llcorner \mathbf{B}(p,r) \xrightarrow{\mathbf{F}} T \llcorner \mathbf{B}(p,r)$ and hence

$$\partial(Q_i \llcorner \mathbf{B}(p,r)) \xrightarrow{\mathbf{F}} \partial(T \llcorner \mathbf{B}(p,r))$$

for almost every r. Recall that by slicing theory 4.11(5), $\partial(Q_i \llcorner \mathbf{B}(p,r))$ and $\partial(T \llcorner \mathbf{B}(p,r))$, and hence of course $Q_i \llcorner \mathbf{B}(p,r)$ and $T \llcorner \mathbf{B}(p,r)$, are normal currents. By induction on (2) above, $Q_i \llcorner \mathbf{B}(p,r)$ and hence $\partial(Q_i \llcorner \mathbf{B}(p,r))$ are integral currents. By induction on (1), $\partial(T \llcorner \mathbf{B}(p,r))$ is an integral current (for almost all r). Now by Lemma 4.12, T is a rectifiable current. Since $\partial T = 0$, T is an integral current, and (1) is proved. As mentioned, (2) and (3) follow.

To prove (4), let T_j be a Cauchy sequence in $\mathscr{T}$. By the completeness of $\{T \in \mathscr{F}_m : \operatorname{spt} T \subset \mathbf{B}(\mathbf{0}, R)\}$, there is a limit $T \in \mathscr{F}_m$. By the lower semicontinuity of mass (Exercise 4.22), $\mathbf{M}(T) \le c$ and $\mathbf{M}(\partial T) \le c$. Finally by (3), $T \in \mathbf{I}_m$.

5.5. The Compactness Theorem [Federer, 4.2.17]. *Let K be a closed ball in $\mathbf{R}^n$, $0 \le c < \infty$. Then*

$$\{T \in \mathbf{I}_m \mathbf{R}^n : \operatorname{spt} T \subset K,\ \mathbf{M}(T) \le c, \text{ and } \mathbf{M}(\partial T) \le c\}$$

is $\mathscr{F}$ compact.

Remark. More generally, K may be a compact C^1 submanifold of $\mathbf{R}^n$ or a compact Lipschitz neighborhood retract (cf. Federer [1, 4.1.29]), yielding the Compactness Theorem in any C^1 compact Riemannian manifold M. (Any C^1 embedding of M in $\mathbf{R}^n$, whether or not isometric, will do, since altering the metric only changes the flat norm by a bounded amount and does not change the topology, as Brian White pointed out to me.)

Proof. The set is complete and totally bounded by the Closure Theorem, 5.4, and the Deformation Theorem Corollary, 5.2.

As an example of the power of the Compactness Theorem, we prove the following corollary.

5.6. The Existence of Area-Minimizing Surfaces. *Let B be an $(m-1)$-dimensional rectifiable current in $\mathbf{R}^n$ with $\partial B = 0$. Then there is an m-dimensional area-minimizing rectifiable current S with $\partial S = B$.*

Remarks. *S area minimizing* means that, for any rectifiable current T with $\partial T = \partial S$, $\mathbf{M}(S) \leq \mathbf{M}(T)$. That B bounds some rectifiable current is shown by taking the cone over B. Even if B is a submanifold, S is not in general.

Proof. Let $\mathbf{B}(\mathbf{0}, A)$ be a large ball containing $\operatorname{spt} B$. Let S_j be a sequence of rectifiable currents with areas decreasing to $\inf\{\mathbf{M}(S) : \partial S = B\}$. The first problem is that the $\operatorname{spt} S_j$ may send tentacles out to infinity. Let Π denote the Lipschitz map which leaves the ball $\mathbf{B}(\mathbf{0}, A)$ fixed and radially projects points outside the ball onto the surface of the ball (Figure 5.6.1). Π is distance nonincreasing and hence area nonincreasing. Therefore by replacing S_j by $\Pi_{\#} S_j$, we may assume that $\operatorname{spt} S_j \subset \mathbf{B}(\mathbf{0}, A)$.

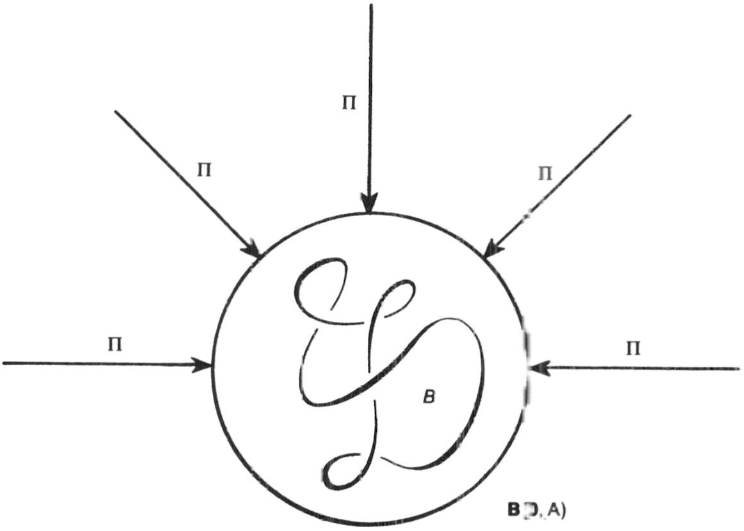

Figure 5.6.1. In the proof of the existence of an area-minimizing surface with given boundary B, it is necessary to keep surfaces under consideration inside some large ball $\mathbf{B}(\mathbf{0}, A)$. This is accomplished by projecting everything outside the ball onto its surface.

Now using the Compactness Theorem we may extract a subsequence which converges to a rectifiable current S. By the continuity of ∂ and the lower semicontinuity of mass (Exercise 4.22), $\partial S = B$ and $\mathbf{M}(S) = \inf\{\mathbf{M}(T) : \partial T = B\}$. Therefore S is the desired area-minimizing surface.

5.7. The Existence of Absolutely and Homologically Minimizing Surfaces in Manifolds [Federer, 5.1.6]. *Let M be a compact, C^1 Riemannian manifold. Let T be a rectifiable current in M. Then among all rectifiable currents S in M such that $\partial S = \partial T$* (respectively, $S - T = \partial X$ *for some rectifiable current X in M*), *there is one of least area.*

S is called absolutely or homologically area minimizing. The methods also treat free boundary problems (cf. 12.3).

Proof. Given the *Remark* after the Compactness Theorem, 5.5, we just need to check that a minimizing limit stays in the same homology class. If $\mathscr{F}(S_i - S)$ is small, $S_i - S = A + \partial B$, with $\mathbf{M}(A)$ and $\mathbf{M}(B)$ small. Let Y_1 be the area minimizer in $\mathbf{R}^n$ with $\partial Y_1 = A$. Since $\mathbf{M}(Y_1)$ is small, by monotonicity 9.5, Y_1 stays close to M and hence may be retracted onto Y in M. Since $S_i - S = \partial Y + \partial B$, therefore $S \sim S_i$, as desired.

EXERCISES

5.1. Verify that the Isoperimetric Inequality, 5.3, is homothetically invariant. (A *homothety* μ_r of $\mathbf{R}^n$ maps x to rx.)

5.2. Prove that 5.4(1) $\Rightarrow$ (2) $\Rightarrow$ (3).

5.3. Try to find a counterexample to the Closure Theorem, 5.4. (Of course, there is none.)

5.4. Show that the Isoperimetric Inequality, 5.3, fails for normal currents.

CHAPTER 6

Examples of Area-Minimizing Surfaces

It can be quite hard to prove that any particular surface is area minimizing. After all, it must compare favorably with all other surfaces with the given boundary. Fortunately, there are a number of beautiful examples.

6.1. The Minimal Surface Equation [Federer, 5.4.18]. *Let f be a C^2, real-valued function on a planar domain D, such that the graph of f is area minimizing. Then f satisfies the minimal surface equation:*

$$\left(1+f_y^2\right)f_{xx}-2f_xf_yf_{xy}+\left(1+f_x^2\right)f_{yy}=0.$$

Conversely, if f satisfies the minimal surface equation on a convex domain, then its graph is area minimizing.

The minimal surface equation just gives the necessary condition that under smooth variations in the surface, the rate of change of the area is 0. This condition turns out to be equivalent to the vanishing of the mean curvature. A smoothly immersed surface which is locally the graph of a solution to the minimal surface equation (or, equivalently, which has mean curvature 0) is called a *minimal surface*. Some famous minimal surfaces are pictured in Figures 6.1.1–6.1.3b.

Theorem 6.1 guarantees that small pieces of minimal surfaces are area minimizing, but larger pieces may not be. For example, the portion of Enneper's surface pictured in Figure 6.1.2 is not area minimizing. There are two area-minimizing surfaces with the same boundary, pictured in

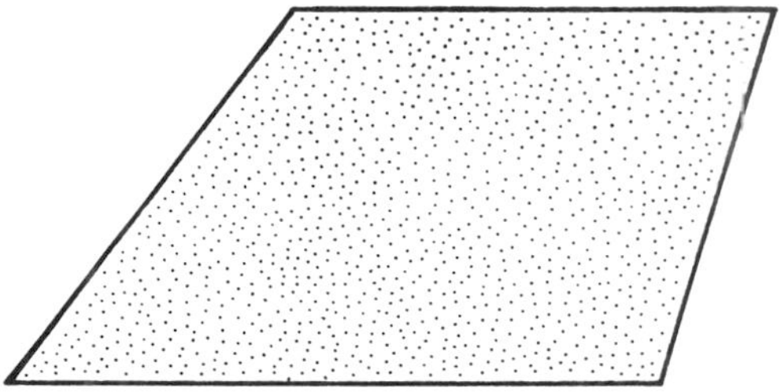

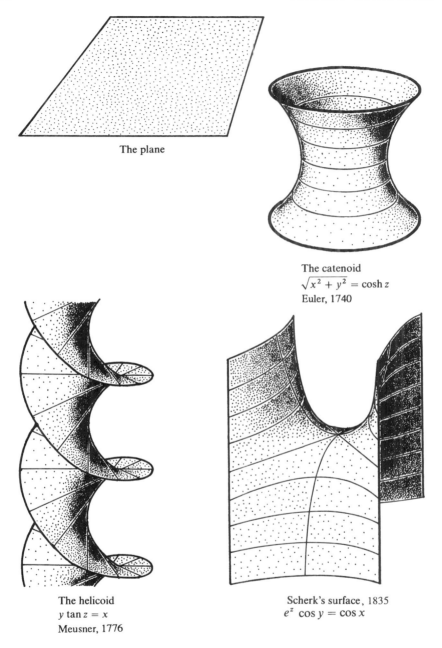

The plane

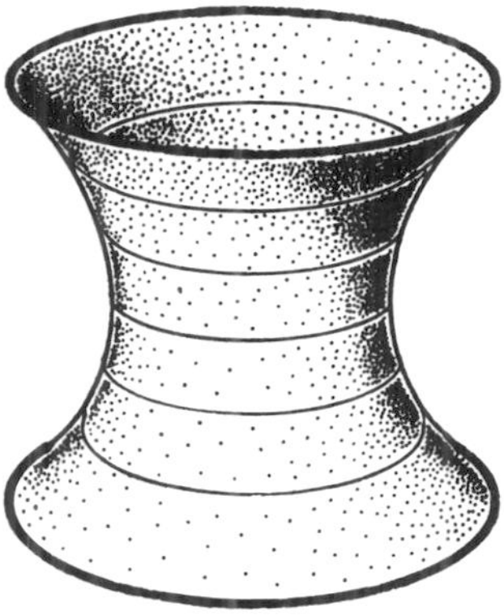

The catenoid
$\sqrt{x^2 + y^2} = \cosh z$
Euler, 1740

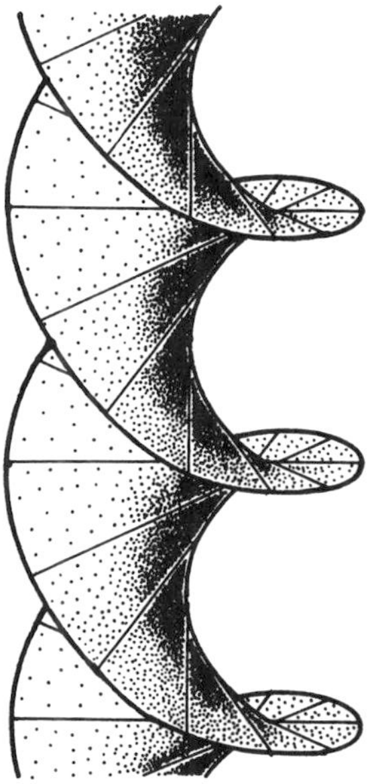

The helicoid
$y \tan z = x$
Meusner, 1776

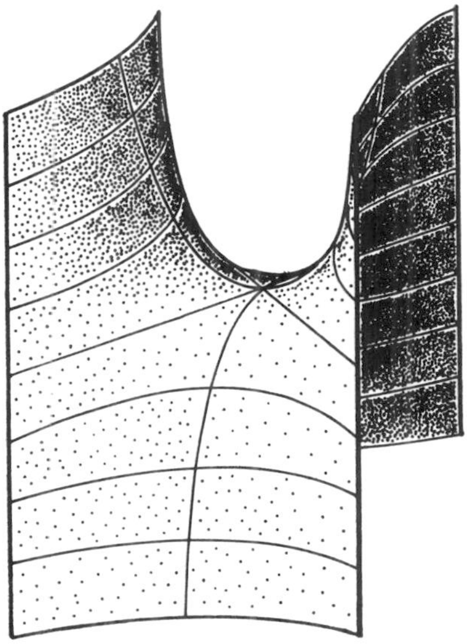

Scherk's surface, 1835
$e^z \cos y = \cos x$

Figure 6.1.1. Some famous minimal surfaces.

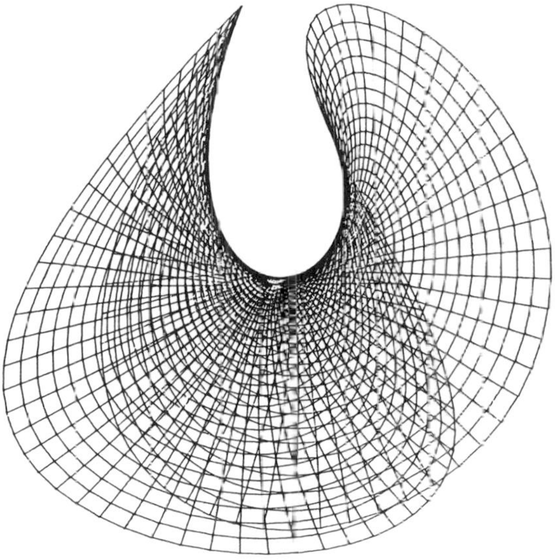

$x = \mathrm{Re}(w - \frac{1}{3}w^3)$
$y = \mathrm{Re}(i(w - \frac{1}{3}w^3))$
$z = \mathrm{Re}(w^2) \qquad w \in \mathbf{C}$

Figure 6.1.2. Enneper's surface, 1864.

Figure 6.1.4. Some systems of curves in $\mathbf{R}^3$ bound infinitely many minimal surfaces. See Morgan [5] and references therein or the popular articles by Morgan [16, 13].

On the disc (or any other convex domain), there is a solution of the minimal surface equation with any given continuous boundary values. We omit the proof. Because the minimal surface equation is not linear, this fact is not at all obvious, and it fails if the domain is not convex. Moreover, on nonconvex domains there can be solutions of the minimal surface equation whose graphs are not area minimizing (Figure 6.1.5).

The first part of the proof of 6.1 will afford an opportunity to illustrate the classical method and notation of the calculus of variations. In general, begin with a function $f\colon D \to \mathbf{R}^m$, supposed to maximize or minimize some functional $A(f)$ for the given boundary values $f|\partial D$. Consider infinitesimal changes δf in f, with $\delta f|\partial D = 0$. Set to 0 the corresponding change in A, called the first variation δA. Use integration by parts to obtain an equation of the form

$$\int_D G(f) \cdot \delta f = 0.$$

a

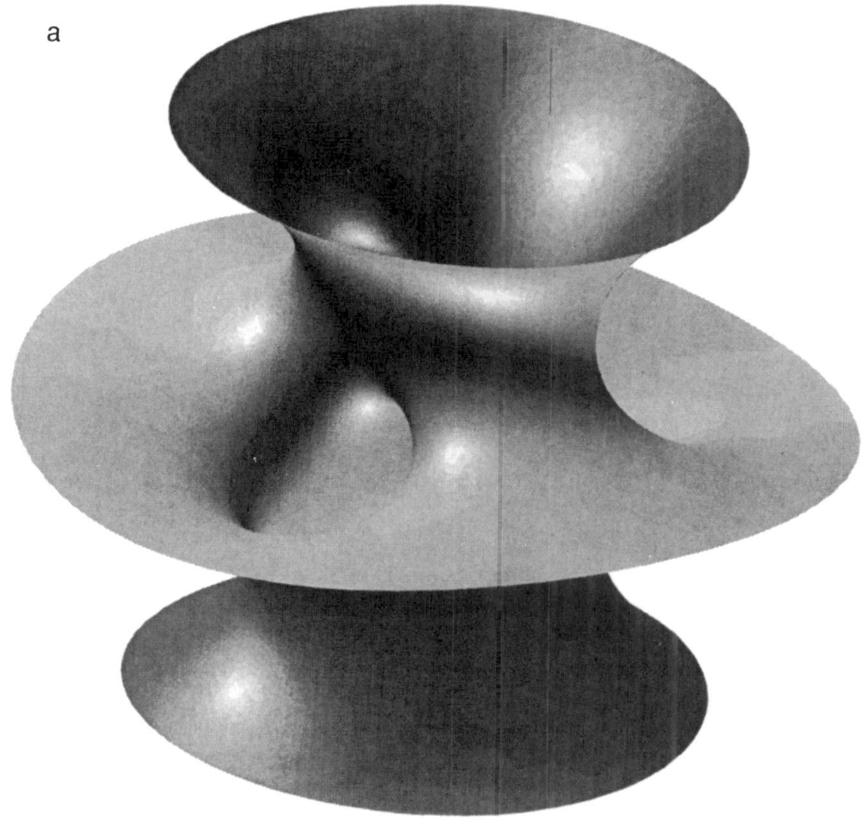

Figure 6.1.3a. The first modern complete, embedded minimal surface of Costa and Hoffman and Meeks (also see Hoffman). Courtesy of David Hoffman, Jim Hoffman, and Michael Callahan.

Since this equation holds for all admissible δf, it follows immediately that $G(f) = 0$. This turns out to be a differential equation for f, called the Euler–Lagrange equation. When $A(f)$ is the area of the graph of f, the associated Euler–Lagrange equation is the minimal surface equation.

Proof of 6.1. First we will show that if the graph of f is area minimizing, then f satisfies the minimal surface equation. By hypothesis, f minimizes the area functional

$$A(f) = \int_D \left(1 + f_x^2 + f_y^2\right)^{1/2} dx\, dy$$

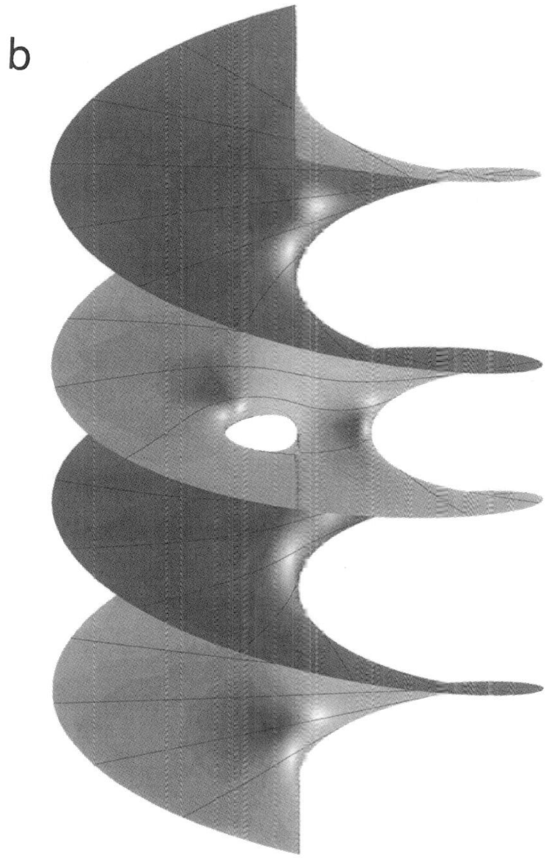

Figure 6.1.3b. One of the latest new complete, embedded minimal surfaces: the genus one helicoid, discovered by David Hoffman, Hermann Karcher, and Fusheng Wei (1993). Computer-generated image made by James T. Hoffman at the GANG Laboratory, University of Massachusetts, Amherst. Copyright GANG, 1993.

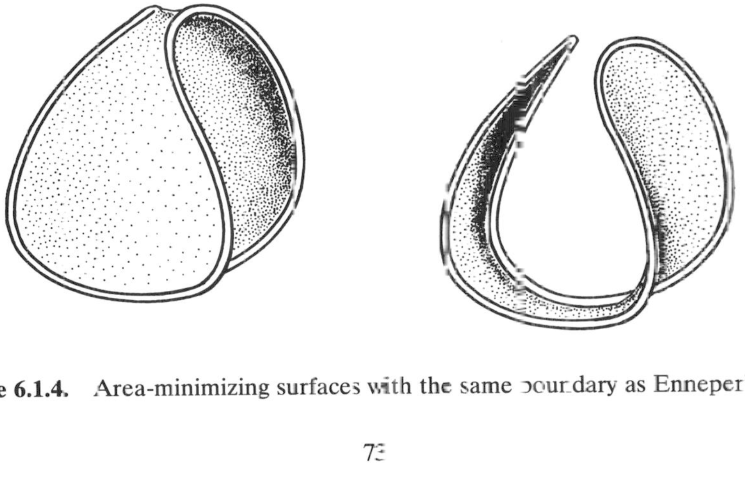

Figure 6.1.4. Area-minimizing surfaces with the same boundary as Enneper's surface.

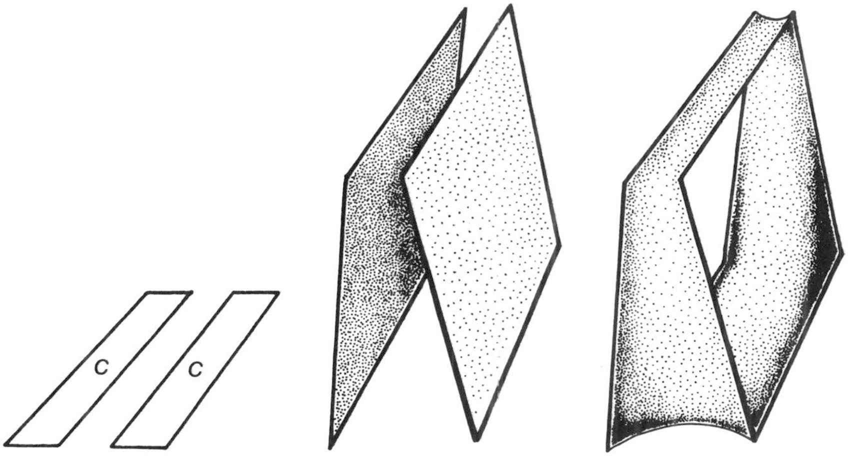

Figure 6.1.5. A minimal graph over a *nonconvex* region C need not be area minimizing. The second surface has less area.

for given boundary values. Therefore, the variation δA in A due to an infinitesimal, smooth variation δf in f, with $\delta f|\partial D = 0$, must vanish:

$$0 = \delta A = \int_D \frac{1}{2}\left(1 + f_x^2 + f_y^2\right)^{-1/2}(2f_x \delta f_x + 2f_y \delta f_y)\,dx\,dy$$

$$= \int_D \left(\left[\left(1 + f_x^2 + f_y^2\right)^{-1/2} f_x\right]\delta f_x + \left[\left(1 + f_x^2 + f_y^2\right)^{-1/2} f_y\right]\delta f_y\right)dx\,dy.$$

Integration by parts now yields an equation of the form $\int_D G(f)\delta f = 0$, where

$$G(f) = -\frac{\partial}{\partial x}\left[\left(1 + f_x^2 + f_y^2\right)^{-1/2} f_x\right] - \frac{\partial}{\partial y}\left[\left(1 + f_x^2 + f_y^2\right)^{-1/2} f_y\right]$$

$$= \frac{1}{2}\left(1 + f_x^2 + f_y^2\right)^{-3/2}(2f_x f_{xx} + 2f_y f_{xy})f_x$$

$$- \left(1 + f_x^2 + f_y^2\right)^{-1/2} f_{xx}$$

$$+ \frac{1}{2}\left(1 + f_x^2 + f_y^2\right)^{-3/2}(2f_x f_{xy} + 2f_y f_{yy})f_y$$

$$- \left(1 + f_x^2 + f_y^2\right)^{-1/2} f_{yy}$$

$$= -\left(1 + f_x^2 + f_y^2\right)^{-3/2}\left[\left(1 + f_y^2\right)f_{xx} - 2f_x f_y f_{xy} + \left(1 + f_x^2\right)f_{yy}\right].$$

Since $\int_D G(f)\delta f = 0$ for all smooth δf satisfying $\delta f|\partial D = 0$, $G(f) = 0$, i.e.,

$$\left(1+f_y^2\right)f_{xx} - 2f_x f_y f_{xy} + \left(1+f_x^2\right)f_{yy} = 0,$$

the minimal surface equation.

Second, we prove that the graph over a convex domain of a solution to the minimal surface equation is area minimizing. This proof uses a *calibration*, i.e., a differential form φ which is closed ($d\varphi = 0$) and has maximum value $\sup \varphi(\xi) = 1$ as a function on the set of unit k-planes ξ. A surface is said to be calibrated by φ if each oriented tangent plane ξ satisfies $\varphi(\xi) = 1$. The method shows that a calibrated surface is automatically area minimizing.

Given $f\colon D \to \mathbf{R}$, define a 2-form φ on $D \times \mathbf{R}$ by

$$\varphi(x, y, z) = \frac{-f_x\, dy\, dz - f_y\, dz\, dx + dx\, dy}{\sqrt{f_x^2 + f_y^2 + 1}}.$$

Then for any point $(x, y, z) \in D \times \mathbf{R}$ and for any unit 2-vector ξ, $\varphi(\xi) \leq 1$, with equality if ξ is tangent to the graph of f at (x, y, z). Moreover, φ is closed:

$$\begin{aligned} d\varphi &= -\frac{\partial}{\partial x} f_x\left(f_x^2 + f_y^2 + 1\right)^{-1/2} - \frac{\partial}{\partial y} f_y\left(f_x^2 + f_y^2 + 1\right)^{-1/2} dx\, dy\, dz \\ &= -\left(f_x^2 + f_y^2 + 1\right)^{-3/2}\left(\left(1+f_y^2\right)f_{xx} - 2f_x f_y f_{xy} + \left(1+f_x^2\right)f_{yy}\right) dx\, dy\, dz \\ &= 0 \end{aligned}$$

by the minimal surface equation.

Now let S denote the graph of f, and let T be any other rectifiable current with the same boundary. Since D is convex, we may assume $\operatorname{spt} T \subset D \times \mathbf{R} = \text{domain } \varphi$, by projecting T into $D \times \mathbf{R}$ if necessary without increasing area T. Now, since $\varphi(\xi) = 1$ whenever ξ is tangent to S,

$$\text{area } S = \int_S \varphi.$$

Since $S - T$ bounds and φ is closed,

$$\int_S \varphi = \int_T \varphi.$$

Since $\varphi(\xi) \le 1$ for all 2-planes ξ,

$$\int_T \varphi \le \operatorname{area} T.$$

Combining the two equations and the inequality yields

$$\operatorname{area} S \le \operatorname{area} T.$$

Therefore S is area minimizing, as desired.

Remark. The same argument with the same calibration shows that if the graph of f has constant mean curvature, then it minimizes area for fixed volume constraint.

6.2. Remarks on Higher Dimensions. For a function $f\colon \mathbf{R}^{n-1} \to \mathbf{R}$, the minimal surface equation takes the form

$$\text{(1)} \qquad \operatorname{div} \frac{\nabla f}{\sqrt{1+|\nabla f|^2}} = 0.$$

The statement and proof of 6.1 apply virtually unchanged.

In higher codimension there is a minimal surface system. For example, for a function $f\colon \mathbf{R}^2 \to \mathbf{R}^{n-2}$, the minimal surface system is

$$\text{(2)} \qquad \left(1+|f_y|^2\right) f_{xx} - 2(f_x \cdot f_y) f_{xy} + \left(1+|f_x|^2\right) f_{yy} = 0.$$

In general codimension, the graph of f need not be area minimizing, even if the domain is convex. (In an attempted generalization of the *Proof* of 6.1, φ generally would not be closed. For a counterexample, see Lawson and Osserman.)

6.3. Complex Analytic Varieties [Federer, 5.4.19]. *Any compact portion of a complex analytic variety in* $\mathbf{C}^n \cong \mathbf{R}^{2n}$ *is area minimizing.*

With this initially astonishing fact, Federer provided some of the first examples of area-minimizing surfaces with singularities. For example, $\{w^2 = z^3\} \subset \mathbf{C}^2$ has an isolated "branch-point" singularity at 0. Similarly, the union of the complex axis planes $\{z = 0\}$ and $\{w = 0\}$ in $\mathbf{C}^2$ has an isolated singularity at 0. These are probably the simplest examples. $\{w = \pm\sqrt{z}\}$ is regular at 0; it is the graph of $z = w^2$. The proof, as that of the second part of 6.1, uses a "calibration" φ.

Proof. On $\mathbf{C}^n = \mathbf{R}^n \oplus \mathbf{R}^n$, let ω be the Kähler form

$$\omega = dx_1 \wedge dy_1 + \cdots + dx_n \wedge dy_n.$$

Wirtinger's Inequality [Federer, 1.8.2, p. 40] says that the real $2p$-form $\varphi = \omega^p / p!$ satisfies

$$|\varphi(\xi)| \leq 1$$

for every $2p$-plane ξ, with equality if and only if ξ is a complex p-plane.

Now let S be a compact portion of a p-dimensional complex analytic variety, and let T be any other $2p$-dimensional rectifiable current with the same boundary. Since S is complex analytic, $\varphi(\xi)$ is 1 on every plane tangent to S, and

$$\text{area}\, S = \int_S \varphi.$$

Since $\partial S = \partial T$, $S - T$ is a boundary. Of course $d\varphi = 0$, because φ has constant coefficients. Therefore

$$\int_S \varphi = \int_T \varphi.$$

Finally, since $\varphi(\xi)$ is always at most 1,

$$\int_T \varphi \leq \text{area}\, T.$$

Combining the equalities and inequality yields

$$\text{area}\, S \leq \text{area}\, T.$$

We conclude that S is area minimizing, as desired.

The second part of the *Proof* of 6.1 and 6.3 proved area minimization by means of a calibration or closed differential form of comass 1, as examples of the following method.

6.4. Fundamental Theorem of Calibrations. *Let φ be a closed differential form of unit comass in $\mathbf{R}^n$ or in any smooth Riemannian manifold M. Let S be an integral current such that $\langle \vec{S}, \varphi \rangle = 1$ at almost all points of S. In $\mathbf{R}^n$, S is area minimizing for its boundary. In any M, S is area minimizing in its homology class (with or without boundary).*

Proof. Let T be any comparison surface. Then

$$\text{area}\, S = \int_S \varphi = \int_T \varphi \leq \text{area}\, T.$$

6.5. History of Calibrations (cf. Morgan [1, 2]). The original example of complex analytic varieties was implicit in Wirtinger (1936), explicit for complex analytic submanifolds in de Rham [1] (1957), and applied to singular complex varieties in the context of rectifiable currents by Federer [3, §4] (1965). Berger [2, §6, last paragraph] (1970) was the first to extract the underlying principle and apply it to other examples such as quaternionic varieties, followed by Dao (1977). The term *calibration* was coined in the landmark paper of Harvey and Lawson, which discovered rich new calibrated geometries of "special Lagrangian," "associative," and "Cayley" varieties.

The method has grown in power and applications. Surveys appear in Morgan [1, 2]. Mackenzie uses calibrations in the proof (Nance; Lawlor [1]) of the angle conjecture on when a pair of m-planes in $\mathbf{R}^n$ is area minimizing. The "vanishing calibrations" of Lawlor [2] actually provide sufficient differential-geometric conditions for area minimization, a classification of all area-minimizing cones over products of m spheres, examples of nonorientable area-minimizing cones, and singularities stable under perturbations. The "paired calibrations" of Lawlor and Morgan and of Brakke [1, 2] and the covering space calibrations of Brakke [3] prove new examples of soap films in $\mathbf{R}^3$, in $\mathbf{R}^4$, and above. Other developments include Murdoch's "twisted calibrations" of nonorientable surfaces, Le's "relative calibrations" of stable surfaces, and Pontryagin calibrations on Grassmannians (Gluck, Mackenzie, and Morgan).

EXERCISES

6.1. Verify that the helicoid is a minimal surface.

6.2. Verify that Scherk's surface is a minimal surface.

6.3. Verify that Enneper's surface is a minimal surface.

6.4. Prove that the catenoids $\sqrt{x^2 + y^2} = \frac{1}{a}\cosh az$ are the only smooth, minimal surfaces of revolution in $\mathbf{R}^3$.

6.5. Verify that for $n = 3$ the minimal surface equation, 6.2(1), reduces to that of 6.1.

6.6. Use 6.2(2) to verify that the complex analytic variety

$$\{w = z^2\} \subset \mathbf{C}^2$$

is a minimal surface.

6.7. Use 6.2(2) to prove that the graph of any complex analytic function $g\colon \mathbf{C} \to \mathbf{C}$ is a minimal surface.

CHAPTER 7

The Approximation Theorem

The Approximation Theorem says that an integral current, T, can be approximated by a slight diffeomorphism of a polyhedral chain, P, or, equivalently, that a slight diffeomorphism $f_{\#}T$ of T can be approximated by P itself. (See Figure 7.1.1.) The approximation P actually coincides with $f_{\#}T$ except for an error term, E, of small mass.

7.1. The Approximation Theorem [Federer, 4.2.20]. *Given an integral current $T \in \mathbf{I}_m\mathbf{R}^n$ and $\varepsilon > 0$, there exist a polyhedral chain $P \in \mathscr{P}_m\mathbf{R}^n$, supported within a distance ε of the support of T, and a C^1 diffeomorphism f of $\mathbf{R}^n$ such that*

$$f_{\#}T = P + E$$

with $\mathbf{M}(E) \le \varepsilon$, $\mathbf{M}(\partial E) \le \varepsilon$, $\operatorname{Lip}(f) \le 1 + \varepsilon$, $\operatorname{Lip}(f^{-1}) \le 1 + \varepsilon$, $|f(x) - x| \le \varepsilon$, *and* $f(x) = x$ *whenever* $\operatorname{dist}(x, \operatorname{spt} T) \ge \varepsilon$.

Proof.

CASE 1. *∂T polyhedral.* Since T is rectifiable, $T = (\mathscr{H}^m \llcorner B) \wedge \xi$, with B rectifiable and $|\xi|$ integer valued. By Proposition 3.11, $\mathscr{H}^m$ almost all of B is contained in a countable union $\bigcup M_i$ of disjoint C^1 embedded manifolds. At almost every point $x \in B$, the density of B and of $\bigcup M_i$ is 1 (Proposition 3.12), so that there is a single M_i such that B and M_i coincide at x except for a set of density 0. Now a covering argument produces a finite collection of disjoint open balls $U_i \subset \mathbf{R}^n - \operatorname{spt} \partial T$ and nearly flat C^1 submanifolds N_i of U_i such that $\bigcup N_i$ coincides with B except for a set of small $\|T\|$ measure. Gentle hammering inside each U_i, smoothed at the edges, is

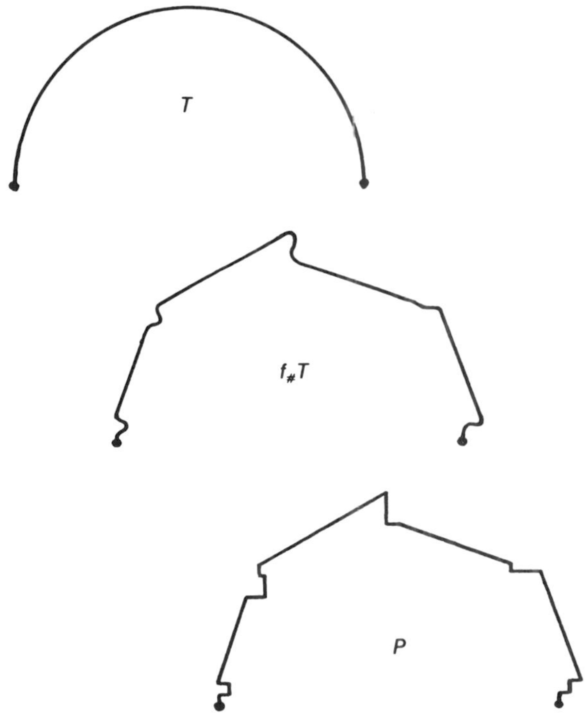

Figure 7.1.1. The Approximation Theorem yields a diffeomorphism $f_\# T$ of T which coincides with a polyhedral P except for small measure.

the desired diffeomorphism f of $\mathbf{R}^n$, which flattens most of B into m-dimensional planes, where $f_\# T$ can be $\mathbf{M}$ approximated by a polyhedral chain P_1 (cf. *Proof* of 4.4).

Unfortunately the error $f_\# T - P_1$, although small in mass, may have huge boundary. Now we use the Deformation Theorem, 5.1, to decompose the error

$$f_\# T - P_1 = P_2 + Q + \partial S.$$

As usual, Q and S have small mass. But in this case, because $f_\# T - P_1$ has small mass, so does P_2 and hence so does the remaining term, ∂S. Because f leaves $\operatorname{spt} \partial T$ fixed, $\partial(f_\# T - P_1) = \partial T - \partial P_1$ is polyhedral, and hence the Deformation Theorem construction makes Q polyhedral. Now take

$$P = P_1 + P_2 + Q.$$

Then $f_\# T - P = \partial S$ has small mass and no boundary.

GENERAL CASE. When ∂T is not polyhedral, first approximate ∂T by the first case,

$$f_{1\#}\partial T = P_1 + \partial S_1,$$

with $\mathbf{M}(S_1)$ and $\mathbf{M}(\partial S_1)$ small. Now $f_{1\#}T - S_1$ has polyhedral boundary and can be approximated by an integral polyhedral chain, P_2,

$$f_{2\#}(f_{1\#}T - S_1) = P_2 + \partial S_2,$$

with $\mathbf{M}(\partial S_2)$ small. Therefore

$$(f_2 \circ f_1)_\# T = P_2 + (f_{2\#}S_1 + \partial S_2),$$

and the error term and its boundary have small mass, as desired.

CHAPTER 8

Survey of Regularity Results

In 1962 Wendell Fleming proved a regularity result that at first sounds too good to be true.

8.1. Theorem [Fleming]. *A two-dimensional, area-minimizing rectifiable current T in $\mathbf{R}^3$ is a smooth, embedded manifold on the interior.*

More precisely, $\operatorname{spt} T - \operatorname{spt} \partial T$ is a C^∞ embedded manifold.

In the classical theory, such complete regularity fails. The disc of least mapping area with given boundary is an immersed minimal surface (see Nitsche [2, §365, p. 318], Osserman [Appendix 3, §1, p. 143], or Lawson [§3, p. 76]), but not in general embedded. For the boundary pictured in Figure 8.1.1, a circle with a tail, the area-minimizing rectifiable current has higher genus, has less area, and is embedded. Pictured in Figure 8.1.2, it flows from the top, flows down the tail, pans out in back onto the disc, flows around front, and flows down the tail to the bottom. There is a hole in the middle that you can stick your finger through. Incidentally, this surface exists as a soap film, whereas the least-area disc does not.

The regularity theorem 8.1 was generalized to three-dimensional surfaces in $\mathbf{R}^4$ by Almgren [6] in 1966 and up through six-dimensional surfaces in $\mathbf{R}^7$ by Simons in 1968. In 1969, Bombieri, De Giorgi, and Giusti gave an example of a seven-dimensional, area-minimizing rectifiable current in $\mathbf{R}^8$ with an isolated interior singularity. Chapter 10 gives a short discussion of this counterexample as well as an outline of the proof of the regularity results.

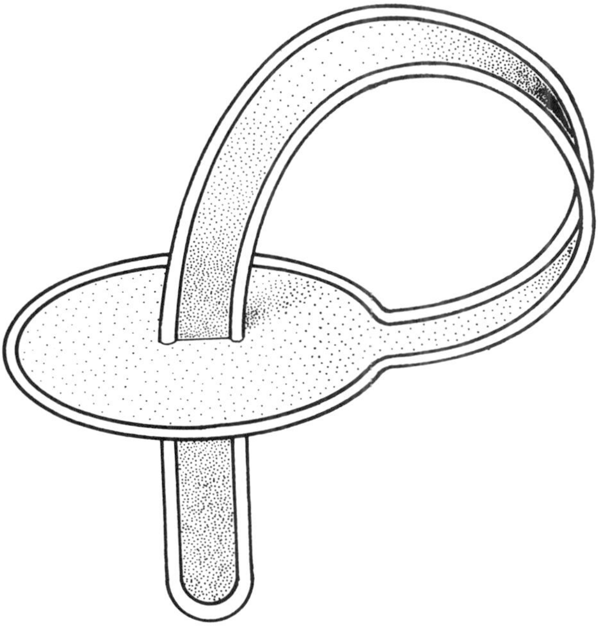

Figure 8.1.1. A least-area disc need not be embedded.

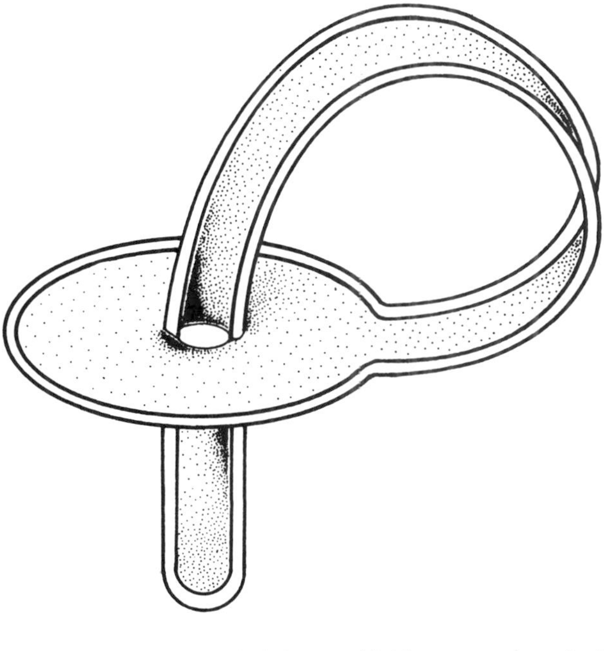

Figure 8.1.2. The area-minimizing rectifiable current is embedded.

The complete interior regularity results for area-minimizing hypersurfaces are given by the following theorem of Federer.

8.2. Theorem [Federer 2]. *An $(n-1)$-dimensional, area-minimizing rectifiable current T in $\mathbf{R}^n$ is a smooth, embedded manifold on the interior except for a singular set of Hausdorff dimension at most $n-8$.*

Regularity in higher codimension, for an m-dimensional area-minimizing rectifiable current T in $\mathbf{R}^n$, with $m < n-1$, is much harder. Until 1983 it was known only that the set of regular points, where $\operatorname{spt} T$ is a smooth embedded manifold, was dense in $\operatorname{spt} T - \operatorname{spt} \partial T$ [Federer, 5.3.16]. On the other hand, m-dimensional complex analytic varieties, which are automatically area minimizing (6.3), can have $(m-2)$-dimensional singular sets. In a major advance Almgren proved the conclusive regularity theorem.

8.3. Theorem [Almgren, 3, 1983]. *An m-dimensional, area-minimizing rectifiable current in $\mathbf{R}^n$ is a smooth, embedded manifold on the interior except for a singular set of Hausdorff dimension at most $m-2$.*

For example, a two-dimensional area-minimizing rectifiable current in $\mathbf{R}^n$ has at worst a zero-dimensional interior singular set. In 1988 Sheldon Chang proved that these singularities must be isolated, "classical branch points."

The stronger regularity theory in codimension 1 comes from an elementary reduction to the relatively easy case of surfaces of multiplicity 1. Indeed, a nesting lemma decomposes an area-minimizing hypersurface into nested, multiplicity-1 area-minimizing surfaces, for which strong regularity results hold. If these surfaces touch, they must coincide by a maximum principle. (See Section 8.5 and Chapter 10.)

In general dimensions and codimensions, very little is known about the structure of the set S of singularities. One might hope that S stratifies into embedded manifolds of various dimensions. However, for all we know to date, S could even be fractional dimensional.

8.4. Boundary Regularity. In 1979 Hardt and Simon proved the conclusive boundary regularity theorem for area-minimizing hypersurfaces.

THEOREM [Hardt and Simon, 1]. *Let T be an $(n-1)$-dimensional, area-minimizing rectifiable current in $\mathbf{R}^n$, bounded by a C^2, oriented sub-*

manifold (with multiplicity 1). *Then at every boundary point*, spt T *is a* C^1, *embedded manifold-with-boundary*.

Notice that general regularity is stronger at the boundary than on the interior, where there can be an $(n-8)$-dimensional singular set. The conclusion is meant to include the possibility that spt T is an embedded manifold without boundary at some boundary points, as occurs for example if the given boundary is two concentric, similarly oriented planar circles. (See Figure 8.4.1.)

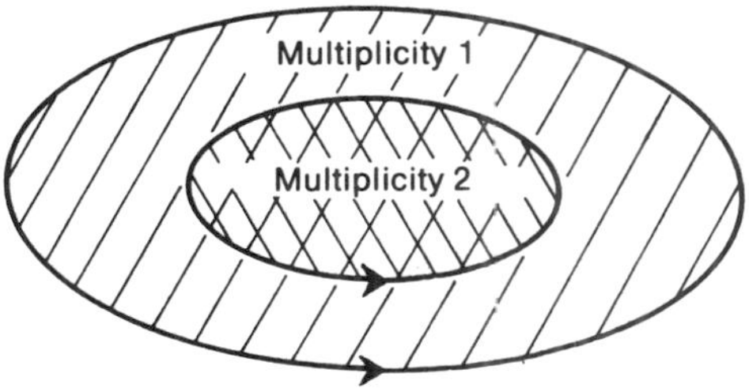

Figure 8.4.1. This example shows that part of the given boundary can end up on the interior of the area-minimizing surface T.

The surface of Figure 8.1.2 at first glance may seem to have a boundary singularity where the tail passes through the disc. Actually, as you move down the tail, the inward conormal rotates smoothly and rapidly almost a full 360°.

B. White [3] has generalized regularity to smooth boundaries with multiplicities.

In higher codimension, boundary singularities can occur.

For the classical least-area disc in $\mathbf{R}^3$, it remains an open question whether there can be boundary branch points if the boundary is C^∞ but not real analytic (see Nitsche [2, §366, p. 320]).

8.5. General Ambients and Other Integrands. All of the above regularity results continue to hold in smooth Riemannian manifolds other than $\mathbf{R}^n$.

Federer [1, 5.4.15, 5.4.4] gives the proof of Theorem 8.2 for $\mathbf{R}^n$ and explains that the methods can be generalized to any smooth Riemannian manifold, except possibly a maximum principle. Because two regular minimal surfaces in local coordinates are graphs of functions satisfying a certain nonlinear, elliptic, partial differential system (see Morgan [6, §2.2] or Federer [1, 5.1.11]), such a maximum principle is standard, due essentially to Hopf [Satz 1′; see also Serrin, p. 184].

Incidentally, Schoen, Simon, and Yau, as well as Schoen and Simon, have proved regularity results for hypersurfaces that are stable but not necessarily area minimizing in Riemannian manifolds of dimension less than eight.

Regularity for surfaces minimizing more general integrands than area is harder, and the results are weaker. For a smooth elliptic integrand Φ (see 12.5) on an n-dimensional smooth Riemannian manifold, an m-dimensional Φ-minimizing rectifiable current is a smooth embedded manifold on an open, dense set [Federer 1, 5.3.17]. If $m = n - 1$, the interior singular set has dimension less than $m - 2$ (Schoen, Simon, and Almgren [II.7, II.9] plus an additional unpublished argument of Almgren as in White [4, Thm. 5.2]). The achievement of such general regularity results is one of the telling strengths of geometric measure theory.

The regularity theory extends to hypersurfaces minimizing area subject to volume constraints.

8.6. Theorem [Gonzalez, Massari, and Tamanini, Theorem 2]. *Let T be an $(n-1)$-dimensional rectifiable current of least area in the unit ball $B \subset \mathbf{R}^n$, with prescribed boundary in ∂B, bounding an oriented region of prescribed volume.*

Then T is a smooth surface of constant mean curvature on the interior, except for a singular set of dimensional most $n-8$.

Gonzalez, Massari, and Tamanini state the result in the equivalent terminology of sets of finite perimeter. Of course once smooth, T must have constant mean curvature by a simple variational argument.

By the methods of Almgren [1], the result also holds in a smooth Riemannian manifold, although there seems to be no clear statement in the literature.

EXERCISES

8.1. Try to come up with a counterexample to Theorems 8.1 and 8.4.

8.2. Try to draw an area-minimizing rectifiable current bounded by the trefoil knot. (Make sure your surface is orientable.)

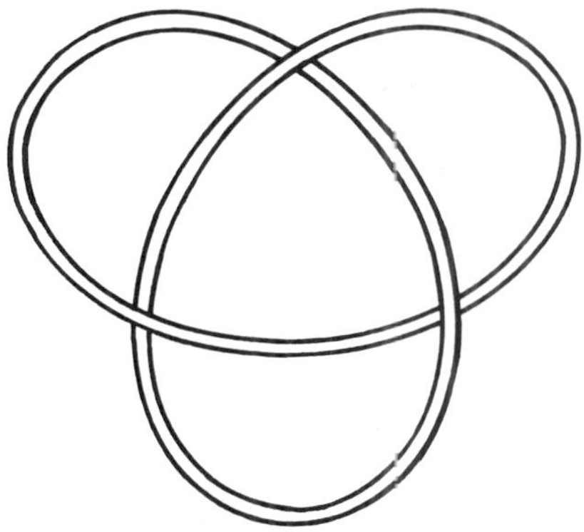

8.3. Illustrate the sharpness of Theorem 8.3 by proving directly that the union of two orthogonal unit discs about 0 in $\mathbf{R}^4$ is area minimizing.
Hint: First show that the area of any surface S is at least the sum of the areas of its projections into the x_1-x_2 and x_3-x_4 planes.

CHAPTER 9

Monotonicity and Oriented Tangent Cones

This chapter introduces the two basic tools of the regularity theory of area-minimizing surfaces. Section 9.2 presents the monotonicity of the mass ratio, a lower bound on area growth. Section 9.3 presents the existence of an oriented tangent cone at *every* interior point of an area-minimizing surface.

9.1. Locally Integral Flat Chains [Federer, 4.1.24, 4.3.16]. We need to generalize our definitions to include noncompact surfaces such as oriented planes. First, define the space $\mathscr{F}_m^{\text{loc}}$ of locally integral flat chains as currents which locally coincide with integral flat chains:

$\mathscr{F}_m^{\text{loc}} = \{T \in \mathscr{D}_m\colon \text{for all } x \in \mathbf{R}^n \text{ there exists } S \in \mathscr{F}_m \text{ with } x \notin \operatorname{spt}(T - S)\}.$

For the local flat topology, a typical neighborhood U_δ of $\mathbf{0}$ takes the form

$$U_\delta = \{T \in \mathscr{F}_m^{\text{loc}}\colon \operatorname{spt}(T - (A + \partial B)) \cap \mathbf{U}(\mathbf{0}, R) = \varnothing,$$
$$A \in \mathscr{R}_m,\ B \in \mathscr{R}_{m+1},\ \mathbf{M}(A) + \mathbf{M}(B) < \delta\},$$

where $\mathbf{U}(\mathbf{0}, R)$ is the open ball $\{x \in \mathbf{R}^n\colon |x| < R\}$.

The subspaces $\mathbf{I}_m^{\text{loc}} \subset \mathscr{R}_m^{\text{loc}} \subset \mathscr{F}_m^{\text{loc}}$ of locally integral currents and locally rectifiable currents are defined analogously:

$\mathbf{I}_m^{\text{loc}} = \{T \in \mathscr{D}_m\colon \text{for all } x \in \mathbf{R}^n \text{ there exists } S \in \mathbf{I}_m \text{ with } x \notin \operatorname{spt}(T - S)\},$

$\mathscr{R}_m^{\text{loc}} = \{T \in \mathscr{D}_m\colon \text{for all } x \in \mathbf{R}^n \text{ there exists } S \in \mathscr{R}_m \text{ with } x \notin \operatorname{spt}(T - S)\}.$

Alternative definitions of the locally rectifiable currents are given by

$$\mathscr{R}_m^{\text{loc}} = \{T \in \mathscr{D}_m\colon T \llcorner \mathbf{B}(\mathbf{0}, R) \in \mathscr{R}_m \text{ for all } R\}$$
$$= \{T \in \mathscr{D}_m\colon T \llcorner \mathbf{B}(a, R) \in \mathscr{R}_m \text{ for all } a \text{ and all } \mathbf{R}\}.$$

There are no similar alternatives for locally integral currents. Indeed, Figure 9.1.1 shows an integral current T such that $T \llcorner \mathbf{B}(\mathbf{0}, 1)$ is not an integral current, because $\mathbf{M}(\partial(T \llcorner \mathbf{B}(\mathbf{0}, 1))) = +\infty$.

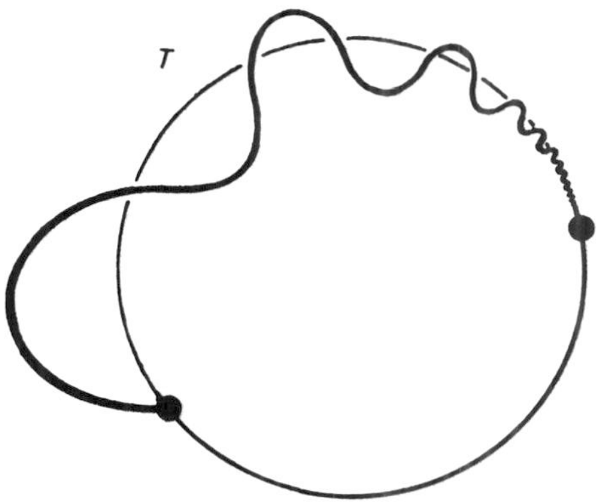

Figure 9.1.1. T is a single curve with two endpoints. However, its restriction to the inside of the circle has infinitely many pieces and infinitely many endpoints. Hence the restriction of an integral current need not be an integral current.

A locally rectifiable current T is called *area minimizing* if for all a and R, $T \llcorner \mathbf{B}(a, R)$ is area minimizing.

The Compactness Theorem, 5.5, generalizes to unbounded locally integral currents in $\mathbf{R}^n$ or noncompact manifolds, as nicely explained in Simon [2, 27.3, 31.2]. Without this perspective, Federer [1] must be forever judiciously restricting currents, a troublesome complication.

9.2. Monotonicity of the Mass Ratio. For a locally rectifiable current $T \in \mathscr{R}_m^{\text{loc}}$ and a point $a \in \mathbf{R}^n$, define the mass ratio

$$\Theta^m(T, a, r) = \mathbf{M}(T \llcorner \mathbf{B}(a, r)) / \alpha_m r^m,$$

where α_m is the measure of the unit ball in $\mathbf{R}^m$. Define the density of T at a,

$$\Theta^m(T, a) = \lim_{r \to 0} \Theta^m(T, a, r).$$

The following theorem on the monotonicity of the mass ratio is one of the most useful tools in regularity theory.

9.3. Theorem [Federer, 5.4.3]. *Let T be an area-minimizing locally rectifiable current in $\mathscr{R}_m^{\text{loc}}$. Let a lie in the support of T. Then for $0 < r <$ $\operatorname{dist}(a, \operatorname{spt} \partial T)$, the mass ratio $\Theta(T, a, r)$ is a monotonically increasing function of r.*

Monotonicity for Minimal Surfaces and Other Integrands. Monotonicity actually holds for stationary (minimal) surfaces or, in a weakened form, even for bounded mean curvature surfaces, but not for more general integrands than area [Allard, 5.1]. (Nevertheless 9.5 holds with a smaller constant for minimizers of general integrands.)

Before giving the Proof of 9.3, we state two immediate corollaries.

9.4. Corollary. *Suppose $T \in \mathscr{R}_m^{\mathrm{loc}}$ is area minimizing. Then $\Theta^m(T, a)$ exists for every $a \in \operatorname{spt} T - \operatorname{spt} \partial T$.*

9.5. Corollary. *Suppose $T \in \mathscr{R}_m^{\mathrm{loc}}$ is area minimizing and $a \in \operatorname{spt} T$. Then for $0 < r < \operatorname{dist}(a, \operatorname{spt} \partial T)$,*

$$\mathbf{M}(T \llcorner \mathbf{B}(a, r)) \geq \Theta^m(T, a) \cdot \alpha_m r^m.$$

For example, if furthermore T happens to be an embedded, two-dimensional, oriented manifold-with-boundary, then

$$\mathbf{M}(T \llcorner \mathbf{B}(a, r)) \geq \pi r^2.$$

See Figures 9.5.1 and 9.5.2.

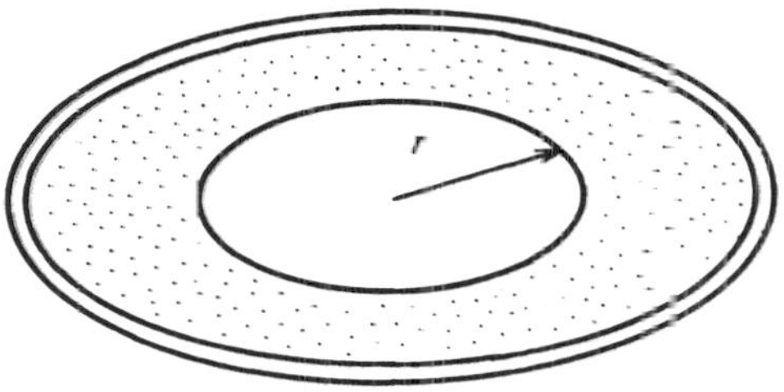

Figure 9.5.1. The disc D is area minimizing and $\mathbf{M}(D \llcorner \mathbf{B}(\mathbf{0}, r)) = \pi r^2$.

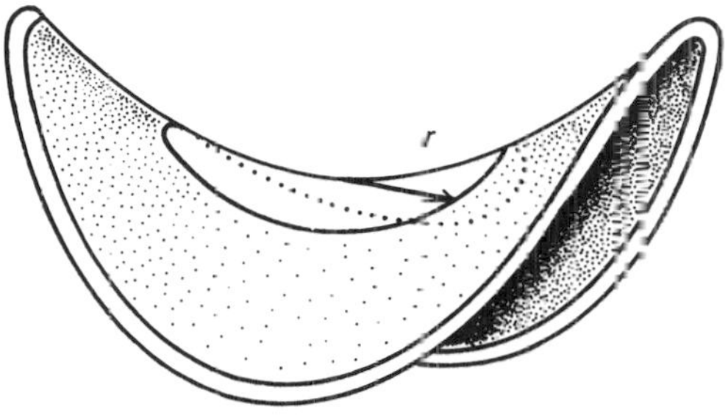

Figure 9.5.2. An area-minimizing surface wtih $\mathbf{M}(T \llcorner \mathbf{B}(\mathbf{0}, r)) > \pi r^2$.

Proof of Theorem 9.3. For $0 < r < \operatorname{dist}(a, \operatorname{spt} \partial T)$, let $f(r)$ denote $\mathbf{M}(T \llcorner \mathbf{B}(a,r))$. Since f is monotonically increasing, for almost all r, $f'(r)$ exists. Slicing by the function $u(x) = |x - a|$ yields (4.11 (3))

$$\mathbf{M}(\partial(T \llcorner \mathbf{B}(a,r))) \leq f'(r).$$

Since T is area minimizing, $\mathbf{M}(T \llcorner \mathbf{B}(a,r))$ is less than or equal to the area of the cone C over $\partial(T \llcorner \mathbf{B}(a,r))$ to a (Figure 9.5.3).

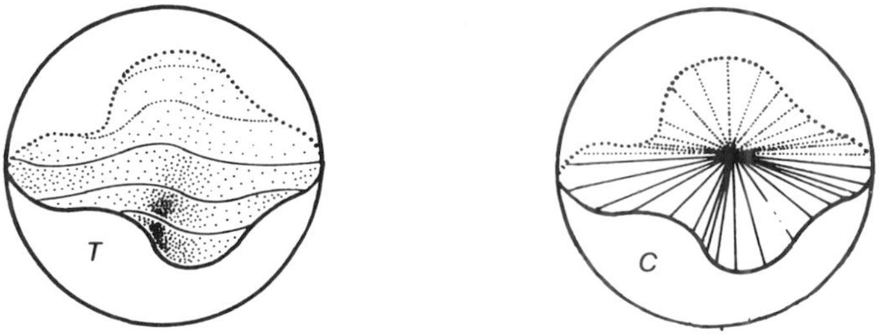

Figure 9.5.3. If $T \llcorner \mathbf{B}(a,r)$ is area minimizing, then it must of course have less area than the cone C over its boundary.

By Exercise 3.8, $\mathbf{M}(C) = \dfrac{r}{m}\mathbf{M}(\partial(T \llcorner \mathbf{B}(a,r)))$. Assembling these inequalities yields

$$f(r) \leq \mathbf{M}(C) = \frac{r}{m}\mathbf{M}(\partial(T \llcorner \mathbf{B}(a,r))) \leq \frac{r}{m} f'(r).$$

Consequently,

$$\begin{aligned}\frac{d}{dr}\alpha_m \Theta(T,a,r) &= \frac{d}{dr}\left[r^{-m} f(r)\right] = r^{-m} f'(r) - m r^{-m-1} f(r)\\ &= \frac{m}{r^{m+1}}\left[\frac{r}{m} f'(r) - f(r)\right] \geq 0.\end{aligned}$$

Hence the absolutely continuous part of $\Theta(T,a,r)$ is increasing. Since any singular part is due to increases in f, $\Theta(T,a,r)$ is increasing as desired.

9.6. Corollary. *Let T be an area-minimizing rectifiable current in $\mathscr{R}_m \mathbf{R}^n$. Then for all $a \in \operatorname{spt} T - \operatorname{spt} \partial T$, $\Theta^m(T,a) \geq 1$.*

Proof. Since a rectifiable set has density 1 almost everywhere (3.12), there is a sequence of points $a_j \to a$ with $\Theta^m(T, a_j) \geq 1$. Let $0 < r <$

$\text{dist}(a, \text{spt}\, \partial T)$, $r_j = \text{dist}(a, a_j)$. Obviously

$$\mathbf{M}(T \llcorner \mathbf{B}(a,r)) \geq \mathbf{M}(T \llcorner \mathbf{B}(a_j, r - r_j)).$$

But by monotonicity, $\mathbf{M}(T \llcorner \mathbf{B}(a_j, r - r_j)) \geq \alpha_m (r - r_j)^m$. Consequently $\mathbf{M}(T \llcorner \mathbf{B}(a,r)) \geq \alpha_m r^m$ and $\Theta(T, a) \geq 1$.

9.7. Oriented Tangent Cones [Federer, 4.3.16]. We now develop a generalization to locally integral flat chains of the notion of the tangent plane to a C^1 manifold at a point. (See Figure 9.7.1.)

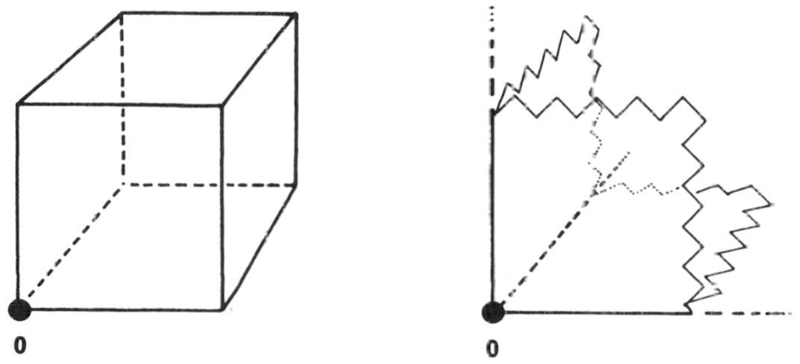

Figure 9.7.1. The surface of the unit cube and the three quarter planes constituting its oriented tangent cone at **0**.

DEFINITIONS. A locally integral flat chain C is called a *cone* if every homothetic expansion or contraction $\mu_{r\#} C = C$. If $T \in \mathscr{F}_m^{\text{loc}}$, such a cone C is called an *oriented tangent cone to* T *at* **0** if there is a decreasing sequence $r_1 > r_2 > r_3 > \ldots$ tending to 0 such that $\mu_{r_j^{-1}\#} T$ converges to C in the local flat topology. Note that an oriented tangent cone C is a current, whereas a tangent cone $\text{Tan}(E, \mathbf{0})$ as defined in Section 3.9 is a set. In general, $\text{spt}\, C \subset \text{Tan}(\text{spt}\, T, \mathbf{0})$, but equality need not hold (cf. Exercise 9.6).

Remarks. Figure 9.7.2 illustrates that an oriented tangent cone is not necessarily unique. As it approaches **0**, this curve alternates between following the x-axis and following the y-axis for successive epochs.

In fact, one of the big open questions in geometric measure theory is whether an area-minimizing rectifiable current T has a unique oriented tangent cone at every point $a \in \text{spt}\, T - \text{spt}\, \partial T$.

Figure 9.7.3 illustrates the need for specifying that C be a cone.

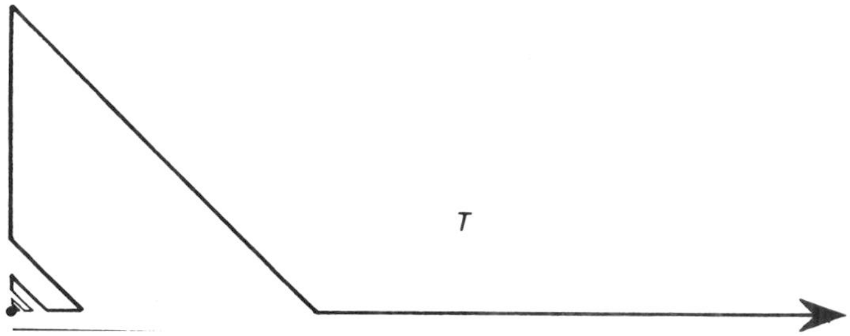

Figure 9.7.2. T alternates ad infinitum between the positive x-axis and the positive y-axis. Each axis is an oriented tangent cone.

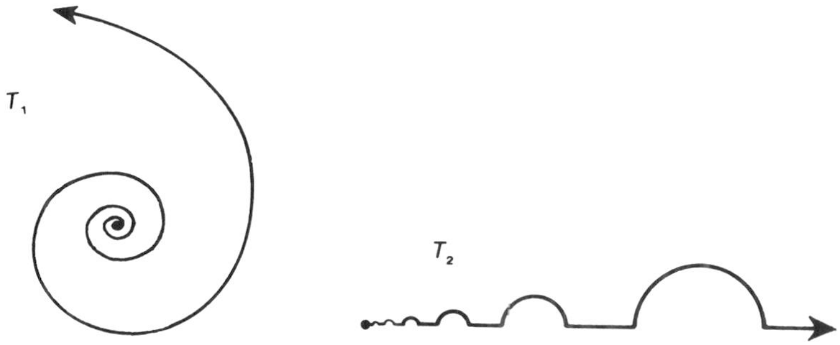

Figure 9.7.3. T_1 and T_2 are both invariant under certain sequences of homothetic expansions, but are not cones.

9.8. Theorem [Federer, 5.4.3(6)]. *Let T be an area-minimizing rectifiable current in $\mathscr{R}_m$. Suppose $\mathbf{0} \in \operatorname{spt} T - \operatorname{spt} \partial T$. Then T has an oriented tangent cone C at $\mathbf{0}$.*

Because the proof is a bit technical, we will prove just this much: there is a rectifiable current C supported in $\mathbf{B}(\mathbf{0}, 1)$ and a sequence $r_1 > r_2 > r_3 > \ldots$ tending to 0 such that the sequence $\mu_{r_j^{-1}\#}(T \llcorner \mathbf{B}(\mathbf{0}, r_j))$ converges to C. The hypotheses of the Compactness Theorem require bounds on both $\mathbf{M}(\mu_{r_j^{-1}\#}(T \llcorner \mathbf{B}(\mathbf{0}, r_j)))$ and $\mathbf{M}(\mu_{r_j^{-1}\#}\partial(T \llcorner \mathbf{B}(\mathbf{0}, r_j)))$. Fix $r_0 < \operatorname{dist}(\mathbf{0}, \operatorname{spt} \partial T)$. Then for $r \leq r_0$, monotonicity of the mass ratio (9.3) says that

$$\begin{aligned} \mathbf{M}(\mu_{r^{-1}\#}(T \llcorner \mathbf{B}(\mathbf{0}, r))) &= \mathbf{M}(T \llcorner \mathbf{B}(\mathbf{0}, r))r^{-m} \\ &\leq \mathbf{M}(T \llcorner \mathbf{B}(\mathbf{0}, r_0))r_0^{-m} \equiv c, \end{aligned}$$

the first desired bound. To get the second bound, slicing theory must be employed to choose the sequence r_j carefully. The theory (4.11(4)) says that for $0 < s < r_0$:

$$\int_{s/2}^{s} \mathbf{M}(\partial(T \llcorner \mathbf{B}(\mathbf{0},r)))\,dr \le \mathbf{M}(T \llcorner \mathbf{B}(\mathbf{0},s)) \le cs^m.$$

Consequently, for some $s/2 < r < s$,

$$\mathbf{M}(\partial(T \llcorner \mathbf{B}(\mathbf{0},r))) \le \frac{cs^m}{s/2} \le 2^m c r^{m-1}$$

and

$$\mathbf{M}(\mu_{r^{-1}\#}\partial(T \llcorner \mathbf{B}(\mathbf{0},r))) \le 2^m c.$$

Therefore a sequence of r_j can be chosen satisfying both desired mass bounds. Now the Compactness Theorem guarantees that some subsequence converges to the desired rectifiable current C, as we set out to prove.

A complete proof of the theorem involves considering not just $\mathbf{B}(\mathbf{0},1)$ but a sequence of balls $\mathbf{B}(\mathbf{0},R_k)$ with $R_k \to \infty$, successively applying the Compactness Theorem to extract subsequences convergent in each $\mathbf{B}(\mathbf{0},R_k)$, and applying a diagonal argument. One difficulty comes in choosing the initial sequence $r_1 > r_2 > r_3 > \dots \to 0$ such that the

$$\overline{\lim_{j\to\infty}}\, \mathbf{M}\left(\mu_{r_j^{-1}\#}\left(T \llcorner \mathbf{B}(\mathbf{0},r_jR_k)\right)\right) < \infty$$

for each k. Another difficulty comes at the end in showing that the limit C is in fact a cone.

9.9. Theorem. *Let T be an area-minimizing rectifiable current in $\mathscr{R}_m$. Suppose $\mathbf{0} \in \operatorname{spt} T - \operatorname{spt} \partial T$. Let C be an oriented tangent cone to T at $\mathbf{0}$. Then* $\Theta^m(C,\mathbf{0}) = \Theta^m(T,\mathbf{0})$.

Remark. Exercise 9.2 implies that C is itself area minimizing.

Proof. After replacing the sequence r_j such that $\mu_{r_j^{-1}\#}T \to C$ with a subsequence if necessary, for each j choose currents A_j, B_j such that

$$spt\left(\mu_{r_j^{-1}\#}T - (A_j + \partial B_j)\right) \cap \mathbf{U}(\mathbf{0},2) = \varnothing,$$

$$\mathbf{M}(A_j) + \mathbf{M}(B_j) \le 1/j^2.$$

Let $u(x) = |x - a|$ and apply slicing theory 4.11(4) to choose $1 < s_j < 1 + 1/j$ such that

$$\mathbf{M}\langle B_j, u, s_j + \rangle \le j\mathbf{M}(B_j) \le 1/j.$$

Note that

$$\mu_{r_j^{-1}\#}T \llcorner \mathbf{B}(\mathbf{0}, s_j) = C \llcorner \mathbf{B}(\mathbf{0}, s_j) + A_j \llcorner \mathbf{B}(\mathbf{0}, s_j) + \partial\big(B_j \llcorner \mathbf{B}(\mathbf{0}, s_j)\big) - \langle \mathbf{B}_j, u, s_j+\rangle.$$

Hence $\mu_{r_j^{-1}\#}T \llcorner \mathbf{B}(\mathbf{0}, s_j) \to C \llcorner \mathbf{B}(\mathbf{0}, 1)$ in the flat norm. By the lower semicontinuity of mass (Exercise 4.22),

$$\Theta(C, \mathbf{0}) \le \Theta(T, \mathbf{0}).$$

Moreover, since $\mu_{r_j^{-1}\#}T\mathbf{B}(\mathbf{0}, s_j)$ is area minimizing and

$$C \llcorner \mathbf{B}(\mathbf{0}, s_j) + A_j \llcorner \mathbf{B}(\mathbf{0}, s_j) - \langle B_j, u, s_j+\rangle$$

has the same boundary,

$$\mathbf{M}\big(\mu_{r_j^{-1}\#}T \llcorner \mathbf{B}(\mathbf{0}, s_j)\big) \le \mathbf{M}\big(C \llcorner \mathbf{B}(\mathbf{0}, s_j)\big) + 2/j.$$

It follows that

$$\Theta(T, \mathbf{0}) \le \Theta(C, \mathbf{0}).$$

EXERCISES

9.1. Give an example of an integral flat chain $T \in \mathscr{F}_0\mathbf{R}^2$ such that $T \llcorner \mathbf{B}^2(\mathbf{0}, 1)$ is not an integral flat chain.

9.2. Let $S_1, S_2, S_3, \ldots \to S$ be a convergent sequence of locally rectifiable currents. Suppose each S_j is area minimizing. Prove that S is area minimizing.

9.3. Let S be an area-minimizing rectifiable current in $\mathscr{R}_2\mathbf{R}^3$ bounded by the circles $x^2 + y^2 = R^2$, $z = \pm 1$ oppositely oriented. Prove that

$$\operatorname{spt} S \subset \left\{\sqrt{x^2 + y^2} \ge R - 2\sqrt{R}\right\}.$$

9.4. Prove or give a counterexample. If $T \in \mathbf{I}_2\mathbf{R}^3$, then for all $a \in \operatorname{spt} T - \operatorname{spt} \partial T$, $\Theta^2(T, a) \ge 1$.

9.5. Let T be an m-dimensional area-minimizing rectifiable current in $\mathbf{R}^n$ and consider

$$f\colon \mathbf{R}^n \to \mathbf{R}, \quad f(x) = \Theta^m(T, x).$$

(a) Mention an example for which f is not continuous, even on $\operatorname{spt} T - \operatorname{spt} \partial T$.

(b) Prove that f is upper semicontinuous on $\mathbf{R}^n - \operatorname{spt} \partial T$.

9.6. Let $T \in \mathscr{F}_m^{\mathrm{loc}}$, and let C be an oriented tangent cone to T at $\mathbf{0}$. Prove that $\operatorname{spt} C \subset \operatorname{Tan}(\operatorname{spt} T, \mathbf{0})$ (cf. 3.9). Show by example that equality need not hold.

9.7. Let $T \in \mathscr{F}_m^{\mathrm{loc}}$, and consider oriented tangent cones to T at $\mathbf{0}$: $C = \lim \mu_{r_j^{-1}\#}T$ and $D = \lim \mu_{s_j^{-1}\#}T$. Prove that, if $0 < \underline{\lim}\, s_j/r_j \le \overline{\lim}\, s_j/r_j < \infty$, then $C = D$.

9.8. Let T be the polygonal curve which follows the x-axis from $1 = 2^{-0^2}$ to 2^{-1^2}, then the y-axis from 2^{-1^2} to 2^{-2^2}, and so on as in Figure 9.7.2, oriented outward.

(a) Show that the nonnegative x-axis and the nonnegative y-axis are each an oriented tangent cone at $\mathbf{0}$.

(b) Find a limit of homothetic expansions which is not a cone.

(c) Let $T_0 = T \cap x$-axis. Show that the lower density $\Theta_*(T_0, \mathbf{0}) = 0$ while the analogous upper density $\Theta^*(T_0, \mathbf{0}) = 1/2$.

CHAPTER 10

The Regularity of Area-Minimizing Hypersurfaces

This chapter outlines some parts of the proof of the regularity theorem for area-minimizing rectifiable currents in $\mathscr{R}_{n-1}\mathbf{R}^n$ for $n \leq 7$. The purpose is to give an overview, illustrate basic arguments, and indicate why regularity fails for $n \geq 8$. The deeper and more technical aspects of the theory are omitted. The first theorem proves a special case by methods that will be useful in the general case.

10.1. Theorem. *Let T be an area-minimizing rectifiable current in $\mathscr{R}_1\mathbf{R}^2$. Then* $\operatorname{spt} T - \operatorname{spt} \partial T$ *consists of disjoint line segments.*

Proof. It will be shown that every point $a \in \operatorname{spt} T - \operatorname{spt} \partial T$ has a neighborhood $\mathbf{U}(a, r)$ such that $\operatorname{spt} T \cap \mathbf{U}(a, r)$ is a straight line segment.

CASE 1. *If ∂T consists of two points (oppositely oriented), then T is the oriented line segment between them.* Our assignment is to prove the most famous result in the calculus of variations: that a straight line is the shortest distance between two points! We may assume $\partial T = \delta_{(1,0)} - \delta_{(0,0)}$. Let T_0 be the oriented segment from $(0,0)$ to $(1,0)$:

$$T_0 = [\![(0,0),(1,0)]\!] = \mathscr{H}^1 \llcorner \{0 \leq x \leq 1, y = 0\} \wedge \mathbf{i}.$$

We will actually show that T_0 uniquely minimizes length among all normal currents $N \in \mathbf{N}_1\mathbf{R}^2$ with the same boundary as T_0. Indeed,

$$\mathbf{M}(N) \geq N(dx) = \partial N(x) = 1 = \mathbf{M}(T_0).$$

Therefore T_0 is area minimizing. Furthermore, if $\mathbf{M}(N)=1$, then $\vec{N}=\mathbf{i}$ $\|N\|$-almost everywhere.

Next, supposing that $\mathbf{M}(N)=1$, we show that $\operatorname{spt} N \subset \{y=0\}$. If not, for some $\varepsilon > 0$ there is a C^∞ function $0 \le f(y) \le 1$ such that $f(y)=1$ for $|y| \le \varepsilon$ and $\mathbf{M}(N \llcorner f) < 1$.

$$\partial(N \llcorner f) = (\partial N) \llcorner f - N \llcorner df = \partial N - 0,$$

because $\vec{N} = \mathbf{i}$ $\|N\|$-almost everywhere and $df(\mathbf{i}) = 0$. Since $\partial(N \llcorner f) = \partial N = \partial T_0$ and T_0 is mass-minimizing, therefore $\mathbf{M}(N \llcorner f) \ge 1$. This contradiction proves that $\operatorname{spt} N \subset \{y=0\}$.

Finally, note that $\partial(N - T_0) = 0$. By the Constancy Theorem 4.9, $N - T_0$ is a multiple of $\mathbf{E}^1 \equiv \mathscr{H}^1 \wedge \mathbf{i}$. Since $N - T_0$ has compact support, it must be 0. Therefore $N = T_0$, uniqueness is proved, and *Case 1* is complete.

CASE 2. *If the density* $\Theta^1(T,a)$ *equals* 1, *then* $\operatorname{spt} T$ *is a straight line segment in some neighborhood* $\mathbf{U}(a,r)$ *of* a. For almost all s, $0 < s < \operatorname{dist}(a, \operatorname{spt} \partial T)$, the slice $\partial(T \llcorner \mathbf{B}(a,s))$ is a zero-dimensional rectifiable current and a boundary, *i.e.*, an even number of points (counting multiplicities). There cannot be 0 points, because then $T' = 0$ would have the same boundary and less mass than $T \llcorner \mathbf{B}(a,s)$. Therefore, $\mathbf{M}(\partial(T \llcorner \mathbf{B}(a,s))) \ge 2$. On the other hand, by slicing theory 4.11(4),

$$s^{-1} \int_0^s \mathbf{M}(\partial(T \llcorner \mathbf{B}(a,r)) \, dr \le s^{-1} \mathbf{M}(T \llcorner \mathbf{B}(a,s)),$$

which converges to $\alpha_1 \Theta^1(T,a) = 2$ as $s \to 0$. Therefore for some small $r > 0$, $\mathbf{M}(\partial(T \llcorner \mathbf{B}(a,r))) = 2$, and $\partial(T \llcorner \mathbf{B}(a,r))$ consists of two points. By *Case 1*, $\operatorname{spt}(T \llcorner \mathbf{B}(a,r))$ is a line segment, as desired.

The general case will require the following lemma.

LEMMA [FEDERER, 4.5.17]. *If* $R \in \mathscr{R}_{n-1} \mathbf{R}^n$ *with* $\partial R = 0$, *then there are nested,* $\mathscr{L}^n$ *measurable sets* $M_i (i \in \mathbf{Z})$, $M_i \subset M_{i-1}$, *such that*

$$R = \sum_{i \in \mathbf{Z}} \partial(\mathbf{E}^n \llcorner M_i) \quad \text{and} \quad \mathbf{M}(R) = \sum_{i \in \mathbf{Z}} \mathbf{M}(\partial(\mathbf{E}^n \llcorner M_i)).$$

Here $\mathbf{E}^n$ is the unit, constant-coefficient n-dimensional current in $\mathbf{R}^n$, defined by

$$\mathbf{E}^n = \mathscr{L}^n \wedge \mathbf{e}_1 \wedge \cdots \wedge \mathbf{e}_n.$$

Proof of Lemma. By the isoperimetric inequality (5.3), $R = \partial T$ for some $T \in \mathbf{I}_n \mathbf{R}^n$. Such a T is of the form $T = \mathbf{E}^n \llcorner f$ for some measurable, integer-valued function f, and $\mathbf{M}(T) = \int |f|$. Just put $M_i = \{x \colon f(x) \ge i\}$. See

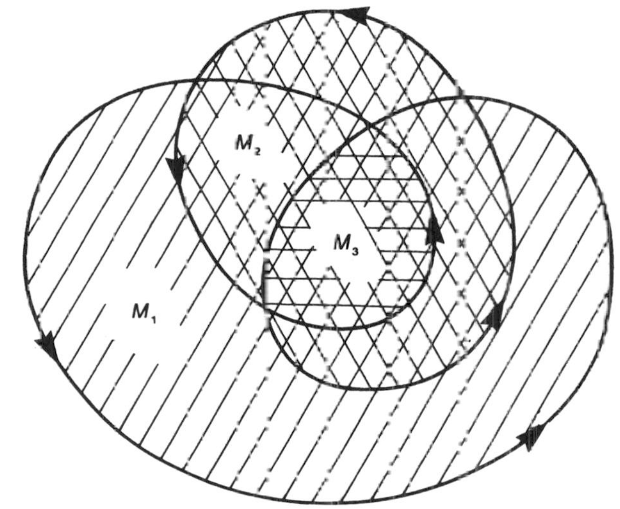

Figure 10.1.1. Decomposing this curve as the boundaries of three nested regions yields three pieces which never cross over each other.

Figure 10.1.1. All of the conclusions of the lemma except the last follow immediately. The last conclusion on $\mathbf{M}(R)$ means that there is no cancellation in the sum $R=\Sigma\partial(\mathbf{E}^n \llcorner M_i)$. The idea of the proof is that, because the M_i are nested, their boundaries, if they happen to overlap, have similar orientations. Hence, in their sum, the masses add. We omit the details.

CASE 3. GENERAL CASE. *For every $a \in \operatorname{spt} T - \operatorname{spt} \partial T$, $\operatorname{spt} T$ is a straight line segment in some neighborhood* $\mathbf{U}(a,r)$ *of* a. Choose $0<\rho<\operatorname{dist}(a, \operatorname{spt}\partial T)$ such that $\mathbf{M}(\partial(T \llcorner \mathbf{B}(a,\rho)))<\infty$. Let Ξ be a rectifiable current supported in the sphere $\mathbf{S}(a,\rho)$ with $\partial\Xi=\partial(T\llcorner\mathbf{B}(a,\rho))$ (Figure 10.1.2). Apply the above lemma to $T\llcorner\mathbf{B}(a,\rho)-\Xi$, and let $T_i=(\partial(\mathbf{E}^2\llcorner M_i))\llcorner\mathbf{U}(a,\rho)$. Since $T\llcorner\mathbf{U}(a,\rho)=\Sigma T_i$ and $\mathbf{M}(T\llcorner\mathbf{U}(a,\rho))=\Sigma\mathbf{M}(T_i)$, each T_i is mass minimizing. Since $\mathbf{M}(T\llcorner\mathbf{U}(a,\rho))<\infty$, it follows from monotonicity (9.5 and 9.6) that $\operatorname{spt} T_i$ intersects $\mathbf{U}(a,\rho/2)$ for only finitely many i, and we will ignore the rest. Since T_i is of the form $(\partial(\mathbf{E}^2\llcorner M_i))\llcorner\mathbf{U}(a,\rho)$, it can be shown with some work that an oriented tangent cone C to T_i at any point in $a\in\operatorname{spt} T_i \cap \mathbf{U}(a,\rho)$ is of the form $C=\partial(\mathbf{E}^2\llcorner N)$ (cf. Federer [1, 5.4.3]). Similarly if $b \in \operatorname{spt} C-\{\mathbf{0}\}$, then an oriented tangent cone D to C at b is of the form $D=\partial(\mathbf{E}^2\llcorner P)$. The fact that C is a cone means that D is a *cylinder* in the sense of invariance under translations in the b direction. A relatively easy argument shows that a one-dimensional oriented cylinder is an oriented

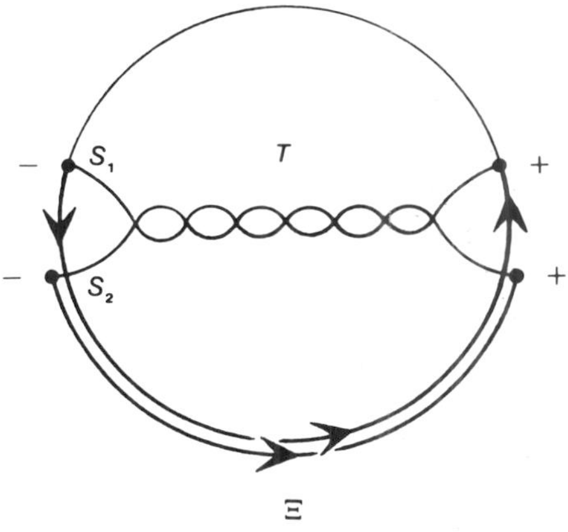

Figure 10.1.2. Given the surface $T \llcorner \mathbf{B}(a,\rho)$ in the ball with boundary in the sphere, there is a surface Ξ entirely in the sphere with the same boundary.

line with multiplicity (cf. Federer [1, 4.3.15]). Since D is of the form $\partial(\mathbf{E}^2 \llcorner P)$, the multiplicity must be 1. Consequently, $\Theta^1(C,b)=\Theta^1(D,\mathbf{0})=1$, for all $b \in \operatorname{spt} C - \{\mathbf{0}\}$. By *Case 2*, C consists of rays emanating from $\mathbf{0}$. Since $C=\partial(\mathbf{E}^2 \llcorner N)$ has no boundary, the same number must be oriented outward as inward. Since C is area minimizing, oppositely oriented rays must be at a 180° angle; otherwise inside the unit circle they could be replaced by a straight line of less mass. See Figure 10.1.3. Therefore C must be an oriented line with multiplicity. Since C is of the form $C=\partial(\mathbf{E}^2 \llcorner N)$, the

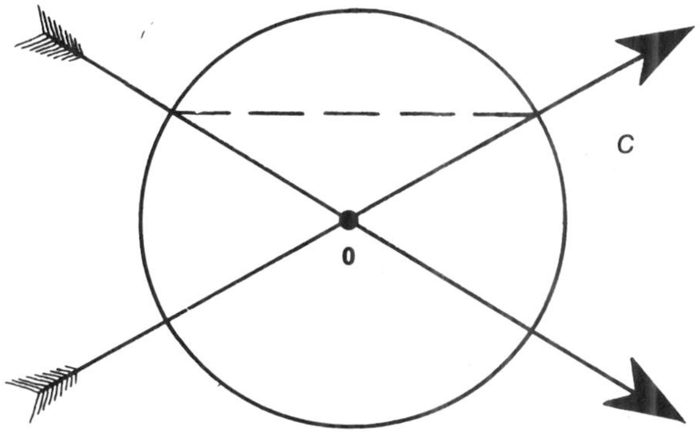

Figure 10.1.3. A pair of lines through $\mathbf{0}$ cannot be mass minimizing, as the dashed shortcut shows.

multiplicity must be 1. Consequently $\Theta^1(T_i, a) = \Theta^1(C, \mathbf{0}) = 1$. By *Case 2*, for some $0 < r_i \le \rho/2$, $T_i \llcorner \mathbf{U}(\mathbf{0}, r_i) = (\partial(\mathbf{E}^2 \llcorner M_i)) \llcorner \mathbf{U}(\mathbf{0}, r_i)$ is an oriented line. Let r be the least of the finitely many r_i. Since $M_i \subset M_{i-1}$, the various nonzero $T_i \llcorner \mathbf{U}(\mathbf{0}, r)$ coincide. Therefore $T \llcorner \mathbf{U}(\mathbf{0}, r)$ is an oriented line with multiplicity.

Now we state and sketch the proof of a complete interior regularity theorem for area-minimizing rectifiable currents in $\mathbf{R}^n$, for $n \le 7$.

10.2. Regularity for Area-Minimizing Hypersurfaces Theorem (Simons; see Federer [1, 5.4.15]). *Let T be an area-minimizing rectifiable current in $\mathscr{R}_{n-1}\mathbf{R}^n$ for $2 \le n \le 7$. Then* $\operatorname{spt} T - \operatorname{spt} \partial T$ *is a smooth, embedded manifold.*

The proof depends on a deep lemma, which we will not prove. Our intent is to show how the pieces fit together. This lemma gives conditions on density and tangent cones sufficient to establish regularity. A later regularity theorem of Allard [Section 8] shows that the hypothesis on an oriented tangent cone is superfluous.

10.3. Lemma [Federer, 5.4.6]. *For $1 \le m \le n-1$, suppose T is an area-minimizing locally rectifiable current in $\mathscr{R}_m^{\mathrm{loc}}\mathbf{R}^n$, $a \in \operatorname{spt} T - \operatorname{spt} \partial T$, $\Theta^m(T, a) = 1$, and some oriented tangent cone to T at a is an oriented, m-dimensional plane. Then* $\operatorname{spt} T$ *is a smooth, embedded manifold at a.*

The second lemma gives a maximum principle for area-minimizing hypersurfaces. The result is contained in Federer [5.3.18], but it is easier to prove.

10.4. Maximum Principle. *For $n \ge 2$, let S_1 and S_2 be $(n-1)$-dimensional, area-minimizing rectifiable currents in $\mathbf{R}^n$, and let $M_i =$* $\operatorname{spt} S_i$. *Suppose that M_1 and M_2 intersect at a point a; that, in some neighborhood of a, M_1 and M_2 are smooth submanifolds; and that M_2 lies on one side of M_1. Then in some neighborhood of a, M_1 and M_2 coincide.*

Proof. At a, M_1 and M_2 may be viewed locally as graphs of functions u_1 and u_2, satisfying the minimal surface equation (6.2(1)). By a standard maximum principle, due essentially to Hopf [Satz 1′], the functions u_1 and u_2 coincide.

The next lemma of Simons provided the final ingredient for the regularity theorem. In a remark following the proof of the theorem we discuss its failure for $n \geq 8$. This lemma marks a solo appearance of differential geometry without measure theory.

10.5. Simons's Lemma [Federer, 5.4.14]. *For $3 \leq n \leq 7$, let B be an oriented, compact, $(n-2)$-dimensional smooth submanifold of the unit $(n-1)$-dimensional sphere, such that the cone over B is area minimizing. Then B is a great sphere.*

It is customarily assumed that the submanifold B is connected. That weaker version of the lemma, applied to the connected components of a general submanifold B, shows that all components are great spheres and hence intersect. This contradiction shows that B is connected and renders that assumption superfluous.

The final lemma has an easy proof by slicing.

10.6. Lemma [Federer, 5.4.8, 5.4.9]. *For $1 \leq m \leq n-1$, let Q be a locally rectifiable current in $\mathscr{R}_m^{\text{loc}}\mathbf{R}^n$. Then Q is area minimizing if and only if $\mathbf{E}^1 \times Q$ is area minimizing.*

Proof of Theorem 10.2. The proof will be in two parts, by induction. The initial case, $n = 2$, was proved by Theorem 10.1.

PART I. *Suppose S is an area-minimizing rectifiable current in $\mathscr{R}_{n-1}\mathbf{R}^n$ and S is of the form $S = (\partial(\mathbf{E}^n \llcorner M)) \llcorner V$ for some measurable set M and open set V. Then* $\operatorname{spt} S \cap V$ *is a smooth embedded manifold.* To prove *Part I*, let $a \in \operatorname{spt} S \cap V$. An oriented tangent cone C to S at a can be shown to be of the form $C = \partial(\mathbf{E}^n \llcorner N)$ (cf. Federer [1, 5.4.3]). Similarly, for $b \in \operatorname{spt} C - \{\mathbf{0}\}$, an oriented tangent cone D to C at b is of the form $D = \partial(\mathbf{E}^n \llcorner P)$. The fact that C is an oriented cone means that D is an "oriented cylinder" of the form $D = \mathbf{E}^1 \times Q$ for some $Q \in \mathscr{R}_{n-2}\mathbf{R}^{n-1}$ (cf. Federer [1, 4.3.15]). Since S is area minimizing, so are C and $D = \mathbf{E}^1 \times Q$. By Lemma 10.6 and induction, Q is an oriented, smooth, embedded manifold, possibly with multiplicity. Hence D is an oriented, smooth, embedded manifold, with multiplicity 1 because D is of the form $D = \partial(\mathbf{E}^n \llcorner P)$. Therefore any oriented tangent cone to D at $\mathbf{0}$ is an oriented, $(n-1)$-dimensional plane. In particular, since D is an oriented cone, D itself is an oriented, $(n-1)$-dimensional plane. Therefore by Lemma 10.3, $\operatorname{spt} C - \{\mathbf{0}\}$ is a smooth, embedded manifold. By Lemma 10.5, $\operatorname{spt} C$ intersects the unit sphere in a great sphere. Hence C is an oriented, $(n-1)$-dimensional plane, with multiplicity 1 because C is of the form $C = \partial(\mathbf{E}^n \llcorner N)$. A reapplication of Lemma 10.3 now shows that spt S is a smooth, embedded manifold at a, proving *Part I*.

PART II. *Completion of proof.* Let $a \in \operatorname{spt} T - \operatorname{spt} \partial T$. Choose a small $\rho > 0$ such that $\partial(T \llcorner \mathbf{U}(a, \rho))$ is rectifiable (cf. 4.11(5)). Let Ξ be a rectifiable current supported in the sphere $\mathbf{S}(a, \rho)$ with the same boundary as $T \llcorner \mathbf{U}(a, \rho)$. By the Lemma of section 10.1, there are nested sets $M_i \subset M_{i-1}$ such that $T \llcorner \mathbf{U}(a, \rho) = \Sigma S_i$, with $S_i = (\partial(\mathbf{E}^n \llcorner M_i)) \llcorner \mathbf{U}(a, \rho)$. Moreover $\mathbf{M}(T \llcorner \mathbf{U}(a, \rho)) = \Sigma \mathbf{M}(S_i)$, so that each S_i is area minimizing. Since $\mathbf{M}(T \llcorner \mathbf{U}(a, \rho)) < \infty$, it follows from monotonicity (9.3 and 9.6) that spt S_i intersects $\mathbf{U}(a, \rho/2)$ for only finitely many i. For such i, by *Part I*, spt S_i is a smooth embedded manifold at a. The containments $M_i \subset M_{i-1}$ imply that each of these manifolds lies on one side of another. Therefore by the maximum principle, 10.4, spt T is a smooth, embedded manifold at a. The theorem is proved.

10.7. Remarks. The regularity theorem fails for $n \geq 8$. E. Bombieri, E. De Giorgi, and E. Giusti (1969), gave an example of a seven-dimensional, area-minimizing rectifiable current T in $\mathbf{R}^8$ with an isolated singularity at $\mathbf{0}$. This current T is the oriented truncated cone over $B = \mathbf{S}^3(\mathbf{0}, 1/\sqrt{2}) \times \mathbf{S}^3(\mathbf{0}, 1/\sqrt{2}) \subset \mathbf{S}^7(\mathbf{0}, 1) \subset \mathbf{R}^8$; $\partial T = B$. It also provides a counterexample to Simons's Lemma 10.5, which is precisely the point at which the proof of regularity breaks down.

Here we give a plausibility argument that such a counterexample should arise for some large n. First consider $B = \mathbf{S}^0(0, 1/\sqrt{2}) \times \mathbf{S}^0(0, 1/\sqrt{2}) \subset \mathbf{S}^1 \subset \mathbf{R}^2$, the four points pictured in Figure 10.7.1. The associated cone does have

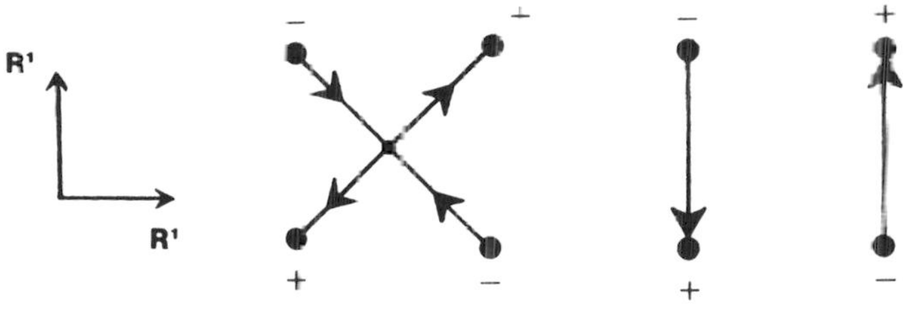

Figure 10.7.1. The X-shape, which is the cone over the four points $\mathbf{S}^0 \times \mathbf{S}^0$ in $\mathbf{R}^2$, is not mass minimizing. The mass-minimizing current consists of two vertical line segments.

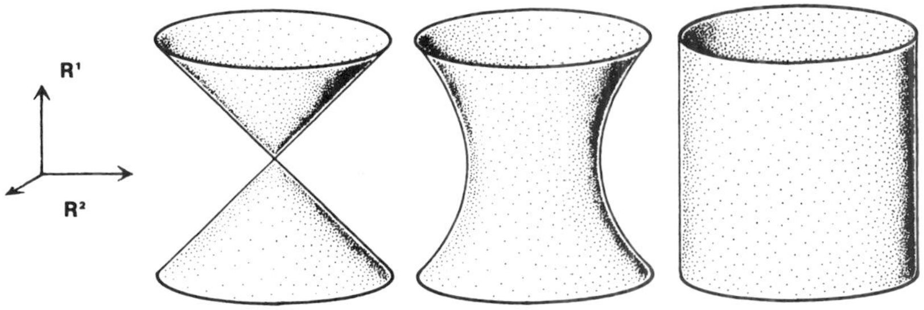

Figure 10.7.2. The cone bounded by the two circles $\mathbf{S}^1 \times \mathbf{S}^0$ in $\mathbf{R}^3$ is not mass minimizing. Neither is the cylinder on the right. The mass-minimizing surface is the catenoid in the middle. It bows inward a bit toward the cone in order to shorten its waist, but not too much to overstretch its sides.

a singularity at $\mathbf{0}$, but it is not mass minimizing. The mass-minimizing current consists of two vertical lines (or two horizontal lines).

Second consider $B = \mathbf{S}^1(\mathbf{0}, 2/\sqrt{5}) \times \mathbf{S}^0(0, 1/\sqrt{5}) \subset \mathbf{S}^2 \subset \mathbf{R}^3$, the two circles pictured in Figure 10.7.2. Again the associated cone has a singularity at $\mathbf{0}$, but it is not mass minimizing. The mass-minimizing current is the pictured catenoid. The catenoid has less area than a cylinder. Although the curved cross-sections are longer than the straight lines of the cylinder, the circumference is less. The amount that the catenoid bows inward toward the cone is a balancing of these two effects.

Third consider $B = \mathbf{S}^2(\mathbf{0}, 1/\sqrt{2}) \times \mathbf{S}^2(\mathbf{0}, 1/\sqrt{2}) \subset \mathbf{S}^5 \subset \mathbf{R}^6$, pictured schematically in Figure 10.7.3. As the dimensions increase, the mass cost of being far from the origin rises, and the mass-minimizing current bows farther toward the cone. Finally in $\mathbf{R}^8$, it has collapsed onto the cone, which is mass minimizing for the first time.

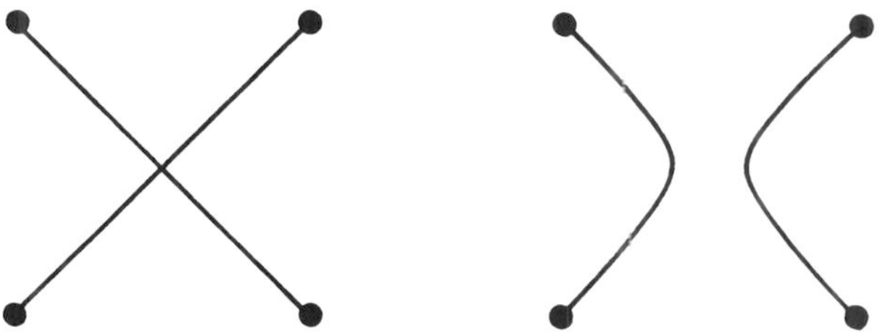

Figure 10.7.3. In higher dimensions, the mass-minimizing surface bows further inward toward the cone. Finally in $\mathbf{R}^8$ it collapses onto the cone, which is mass minimizing for the first time.

EXERCISES

10.1. Give two different area-minimizing rectifiable currents in $\mathscr{R}_2\mathbf{R}^3$ with the same boundary.

10.2. Prove or give a counterexample: If $T \in \mathscr{R}_m(\mathbf{R}^n \times \mathbf{R}^l)$ is area minimizing and p denotes orthogonal projection of $\mathbf{R}^n \times \mathbf{R}^l$ onto $\mathbf{R}^n$, then $p_\# T$ is area minimizing.

10.3. Find a counterexample in $\mathbf{I}_2\mathbf{R}^3$ to Lemma 10.3 if the hypothesis that S be area minimizing is removed.

CHAPTER 11

Flat Chains Modulo ν, Varifolds, and $(\mathbf{M}, \varepsilon, \delta)$-Minimal Sets

A number of alternative spaces of surfaces have been developed in geometric measure theory, as required for theory and applications. This chapter gives brief descriptions of flat chains modulo ν, varifolds, and $(\mathbf{M}, \varepsilon, \delta)$-minimal sets.

11.1. Flat Chains Modulo ν [Federer, 4.2.26]. One way to treat nonorientable surfaces and more general surfaces is to work modulo 2, or more generally modulo ν for any integer $\nu \geq 2$. Two rectifiable currents T_1, T_2 are *congruent modulo* ν if $T_1 - T_2 = \nu Q$ for some rectifiable current Q. In particular, $T \equiv -T \pmod 2$. The m-dimensional rectifiable currents modulo ν, denoted $\mathscr{R}_m^\nu$, are defined as congruence classes of rectifiable currents.

For example, consider the Möbius strip of Figure 11.1.2 bounded by the curve C. There is no way to orient it to turn it into a rectifiable current with boundary C. However, if it is cut along a suitable curve D, it can then be oriented as a rectifiable current T, and it works out that $\partial T \equiv C \pmod 2$. In general, rectifiable currents modulo 2 correspond to unoriented surfaces.

Two parallel circles, as in Figure 11.1.3, bound an interesting rectifiable current modulo 3. The surfaces of both Figures 11.1.2 and 11.1.3 occur as soap films.

Most of the concepts and theorems on rectifiable currents have analogs for rectifiable currents modulo ν: mass, flat norm, the Deformation

Figure 11.1.1. Bill Ziemer (right), who introduced flat chains modulo 2, with his thesis advisor, Wendell Fleming (left), and the author (center), at a celebration in Ziemer's honor at Indiana in 1994. Photo courtesy of Ziemer.

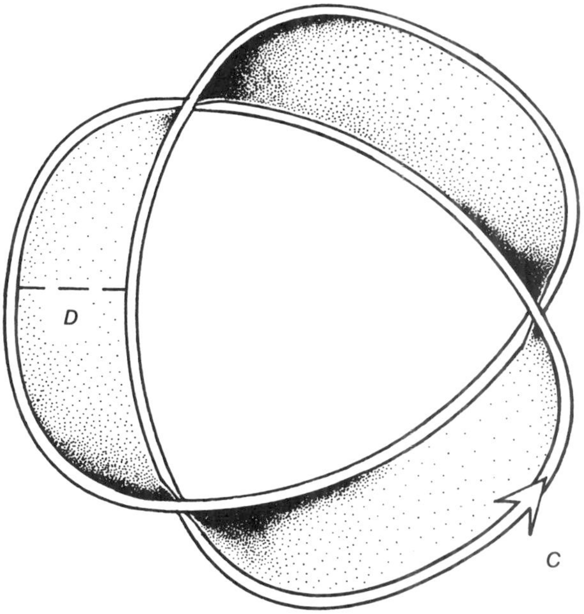

Figure 11.1.2. A rectifiable current modulo 2 can be nonorientable, like this Möbius strip.

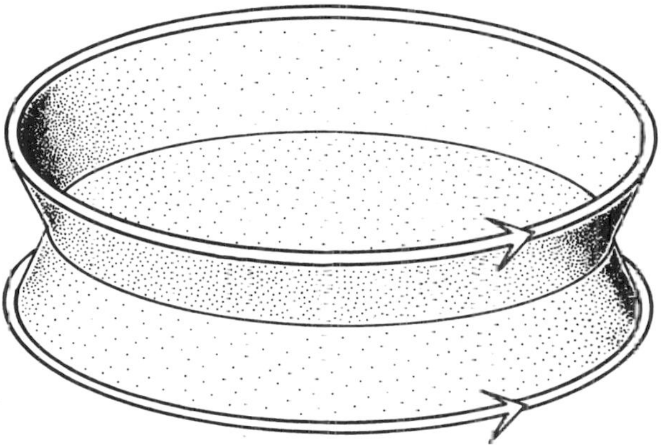

Figure 11.1.3. In a rectifiable current modulo 3, three sheets can meet along a curve inside the surface which does not count as boundary.

Theorem, the Compactness Theorem, existence theory, and the Approximation Theorem.

One tricky point: to ensure completeness, the integral flat chains congruent to 0 modulo ν must be defined as the *flat-norm closure* of those of the form νQ. In codimension greater than 1, it is an open question whether they are all of the form νQ. (The counterexample in Federer [p. 426] is wrong.) However, every rectifiable current congruent to 0 modulo ν is of the form νQ.

It is generally easier to prove regularity for area-minimizing rectifiable currents modulo 2 than for rectifiable currents themselves. Indeed, an m-dimensional area-minimizing rectifiable current modulo 2 in $\mathbf{R}^n$ is a smooth embedded manifold on the interior, except for a rectifiable singular set of locally finite $\mathscr{H}^{m-2}$ measure (Federer [2]; Simon [1]). The only standard regularity result I know which fails modulo 2 is boundary regularity for area-minimizing hypersurfaces (8.4).

For $\nu > 2$, even area-minimizing hypersurfaces modulo ν can have codimension 1 singular sets, as Figure 11.1.3 suggests. Taylor [3] proved that a two-dimensional area-minimizing rectifiable current modulo 3 in $\mathbf{R}^3$ away from boundary consists of C^∞ surfaces meeting in threes at 120° angles along $C^{1,\alpha}$ curves (cf. 13.9). White [5] proved that an $(n-1)$-dimensional area-minimizing rectifiable current modulo 4 in $\mathbf{R}^n$ decomposes locally into a pair of area-minimizing rectifiable currents modulo 2. Moreover, White

[4] proved that for any odd ν, an $(n-1)$-dimensional area-minimizing rectifiable current modulo ν in $\mathbf{R}^n$ away from boundary is a smooth embedded manifold, except for a singular set of dimension at most $n-2$. For even $\nu > 2$, almost everywhere regularity remains an open question.

11.2. Varifolds [Allard]. Varifolds provide an alternative perspective to currents for working with rectifiable sets. Varifolds carry no orientation, and hence there is no cancellation in the limit and no obvious definition of boundary.

An m-dimensional *varifold* is a Radon measure on $\mathbf{R}^n \times G_m\mathbf{R}^n$, where $G_m\mathbf{R}^n$ is the Grassmannian of unoriented unit m-planes through $\mathbf{0}$ in $\mathbf{R}^n$. We have seen previously how to associate to a rectifiable set E the measure $\mathscr{H}^m \llcorner E$ on $\mathbf{R}^n$, but this perspective ignores the tangent planes to E. Instead, we now associate to a rectifiable set E, with unoriented tangent planes $\vec{E}(x) = \mathrm{Tan}^m(E, x) \in G_m\mathbf{R}^n$, the varifold

$$\mathbf{v}(E) \equiv \mathscr{H}^m \llcorner \left\{ \left(x, \vec{E}(x) \right) : x \in E \right\}.$$

The varifolds which so arise are called *integral varifolds*. One allows positive integer multiplicities and noncompact support.

In general varifolds, the tangent planes need not be associated with the underlying set. For example, if you cut an edge off a cubical crystal, as in Figure 11.2.1, because the crystal loves horizontal and vertical directions, the exposed surface forms very small horizontal and vertical steps. Such a corrugation can be modeled by a varifold concentrated half and half on horizontal and vertical tangent planes in $G_2\mathbf{R}^3$, whereas in space $\mathbf{R}^3$ it is concentrated on the diagonal surface.

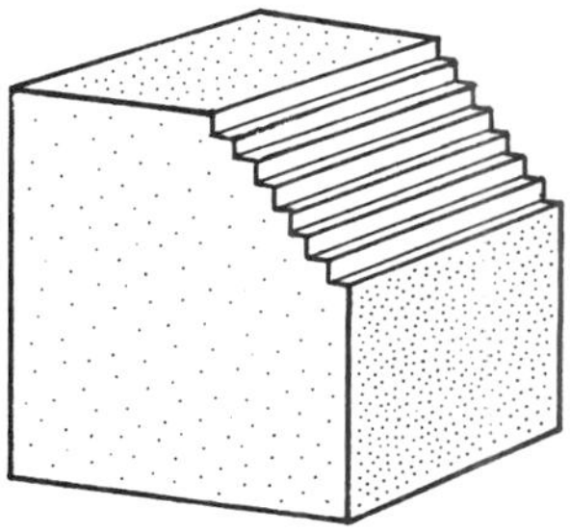

Figure 11.2.1. If you cut an edge off a cubical crystal, the exposed surface forms tiny steps, well modeled by a varifold.

The *first variation* δV of a varifold V is a function which assigns to any compactly supported smooth vectorfield g on $\mathbf{R}^n$ the initial rate of change of the area of V under a smooth deformation of $\mathbf{R}^n$ with initial velocity g. Roughly, the first variation is due to the mean curvature of V and the boundary of V. A varifold is called *stationary* if $\delta V = 0$. Geometrically, stationary integral varifolds include area-minimizing rectifiable currents, area-minimizing rectifiable currents modulo ν, and many other physical surfaces such as soap films. Some singularities of soap films unavoidably count as additional boundary in the category of rectifiable currents or rectifiable currents modulo ν, but fortunately do not add to their first variations as varifolds.

There is a compactness theorem for integral varifolds with bounds on their areas, first variations, and supports. There are also general isoperimetric and regularity theorems. However, it is an open question whether a two-dimensional stationary integral varifold in an open subset of $\mathbf{R}^3$ is a smooth embedded manifold almost everywhere.

11.3. $(\mathbf{M}, \varepsilon, \delta)$-Minimal Sets [Almgren 1]. Perhaps the best model of soap films is provided by the $(\mathbf{M}, 0, \delta)$-minimal sets of Almgren. A nonempty, bounded subset $S \subset \mathbf{R}^n - B$ with $\mathscr{H}^m(S) < \infty$ and $S = \operatorname{spt}(\mathscr{H}^m \llcorner S) - B$ is $(\mathbf{M}, 0, \delta)$ *minimal* with respect to a closed set B (typically "the boundary") if, for every Lipschitz deformation φ of $\mathbf{R}^n$ which differs from the identity map only in a δ-ball disjoint from B,

$$\mathscr{H}^m(S) \leq \mathscr{H}^m(\varphi(S)).$$

Since φ need not be a diffeomorphism, it can pinch pieces of surface together, as in Figure 11.3.1. The $\mathbf{M}$ refers to area and may be replaced by a

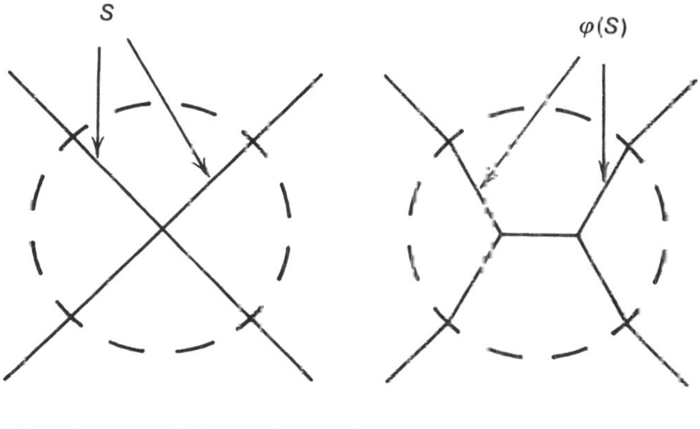

Figure 11.3.1. A curve S is not $(\mathbf{M}, 0, \delta)$ minimal if a deformation $\varphi(S)$ has less length.

more general integrand. For more general functions $\varepsilon(r) = Cr^{\alpha}$, $\alpha > 0$, the inequality imposed on deformations inside r-balls ($r \le \delta$) relaxes to

$$\mathscr{H}^m(S) \le (1 + \varepsilon(r))\mathscr{H}^m(\varphi(S)).$$

Such $(\mathbf{M}, \varepsilon, \delta)$-minimal sets include soap bubbles (with volume constraints); see Section 13.8.

Three basic properties of $(\mathbf{M}, \varepsilon, \delta)$-minimal sets are $(\mathscr{H}^m, m)$ rectifiability [Almgren 1, II.3(9)], monotonicity [Taylor 4, II.1], and the existence of an $(\mathbf{M}, 0, \delta)$-minimal tangent cone at every point [Taylor 4, II.2]. The condition in the definition that $S = \operatorname{spt}(\mathscr{H}^m \llcorner S) - B$ may be replaced by the condition that S be rectifiable, with the understanding that S may be altered by a set of $\mathscr{H}^m$ measure 0 [Morgan 12, §2.5].

Almgren [1] has proved almost everywhere regularity results for $(\mathbf{M}, \varepsilon, \delta)$-minimal sets (except for the special one-dimensional case, treated in Morgan [7]). In 1976, Taylor [4] proved that for two-dimensional $(\mathbf{M}, \varepsilon, \delta)$-minimal sets in $\mathbf{R}^3$, there are only two possible kinds of singularities: (1) three sheets of surface meeting at 120° angles along a curve and (2) four such curves meeting at approximately 109° angles at a point (see Section 13.9). These are precisely the two kinds of singularities that Plateau had observed in soap bubbles and soap films a hundred years earlier. (Warning: Almgren [1] and Taylor [4] are technical, and I think there are some (correctable) gaps.)

Boundary regularity remains conjectural; see Morgan [18, Lecture IV].

It remains an open question today whether a smooth curve in $\mathbf{R}^3$ bounds a least-area soap film in the class of $(\mathbf{M}, 0, \delta)$-minimal sets, with variable $\delta > 0$. The problem is that in the limit δ may go to 0. See Morgan [15]. Another open question asks whether the Cartesian product of an $(\mathbf{M}, 0, \delta)$-minimal set with an interval is $(\mathbf{M}, 0, \delta')$ minimal (a property which holds trivially for most classes of minimal surfaces; cf. 10.6).

EXERCISES

11.1. Give an example of a boundary curve in $\mathbf{R}^3$ for which the area-minimizing flat chain modulo 4 has less area than the area-minimizing integral current.

11.2. Let S be the unit 2 disc, and let $\mathbf{v}(S)$ be the associated varifold. What is $\mathbf{v}(S)(\mathbf{R}^n \times G_2\mathbf{R}^n)$?

11.3. Give an example of a two-dimensional set in $\mathbf{R}^3$ that is $(\mathbf{M}, 0, \delta)$ minimal for small δ but not for large δ.

11.4. Give an example of a two-dimensional set in $\mathbf{R}^3$ that is $(\mathbf{M}, 0, \delta)$ minimal for all $\delta > 0$ but not area minimizing.

CHAPTER 12

Miscellaneous Useful Results

Federer's treatise presents many basic methods of geometry and analysis in a generality that embraces manifold applications. This chapter describes Federer's treatment of Sard's Theorem, Green's Theorem, relative homology, and functions of bounded variation.

12.1. Morse–Sard–Federer Theorem. The usual statement of Sard's Theorem says that the set of critical values of a C^∞ function $f\colon \mathbf{R}^m \to \mathbf{R}^n$ has Lebesgue measure 0. Federer's refinement shows precisely how the Hausdorff measure of the image depends on the rank of Df and the smoothness class of f.

THEOREM [FEDERER, 3.4.3]. *For integers $m > \nu \geq 0$, $k \geq 1$, let f be a C^k function from an open subset A of $\mathbf{R}^m$ into a normed vectorspace Y. Then*

$$\mathscr{H}^{\nu + \frac{m-\nu}{k}} f(\{x \in A\colon \operatorname{rank} Df(x) \leq \nu\}) = 0.$$

Note that the usual statement may be recovered by taking $Y = \mathbf{R}^n$, $\nu = n-1$, $k \geq m - n + 1$. The latest improvement has been provided by Bates.

12.2. Gauss–Green–De Giorgi–Federer Theorem. The usual statement of Green's Theorem says that a C^1 vectorfield $\xi(x)$ on a compact region A in $\mathbf{R}^n$ with C^1 boundary B satisfies

$$\int_B \xi(x) \cdot \mathbf{n}(A, x)\, d\sigma = \int_A \operatorname{div} \xi(x)\, d\mathscr{L}^n x,$$

where $\mathbf{n}(A, x)$ is the exterior unit normal to A at x and $d\sigma$ is the element of area on B. Federer treats more general regions and vectorfields that are merely Lipschitz.

Federer allows measurable regions A for which the current boundary $T = \partial(\mathbf{E}^n \llcorner A)$ is representable by integration: $T(\varphi) = \int \langle \vec{T}, \varphi \rangle d\|T\|$ (cf. 4.3). If A is compact, this condition just says that the current boundary has finite measure: $\mathbf{M}(T) = \|T\|(\mathbf{R}^n) < \infty$. In any case, this condition is weaker than requiring that the topological boundary of A have finite $\mathscr{H}^{n-1}$ measure.

Definition. Let $b \in A \subset \mathbf{R}^n$. We call $\mathbf{n} = \mathbf{n}(A, b)$ the *exterior normal of* A *at* b if $\mathbf{n}$ is a unit vector,

$$\Theta^n(\{x : (x - b) \cdot \mathbf{n} > 0\} \cap A, b) = 0,$$

and

$$\Theta^n(\{x : (x - b) \cdot \mathbf{n} < 0\} - A, b) = 0.$$

Clearly there is at most one such $\mathbf{n}$. If b is a smooth boundary point of A, then $\mathbf{n}$ is the usual exterior normal. Even if ∂A is not smooth at b, $\mathbf{n}$ may be defined, as Figure 12.2.1 suggests. The assertion in the theorem below that the measure $\|T\| = \mathscr{H}^{n-1} \llcorner \operatorname{domain} \mathbf{n}(A, x)$ says roughly that the current boundary of A coincides with the domain of $\mathbf{n}(A, x)$ almost everywhere. In the final formula, $\operatorname{div} \xi$ exists almost everywhere because a Lipschitz function is differentiable almost everywhere.

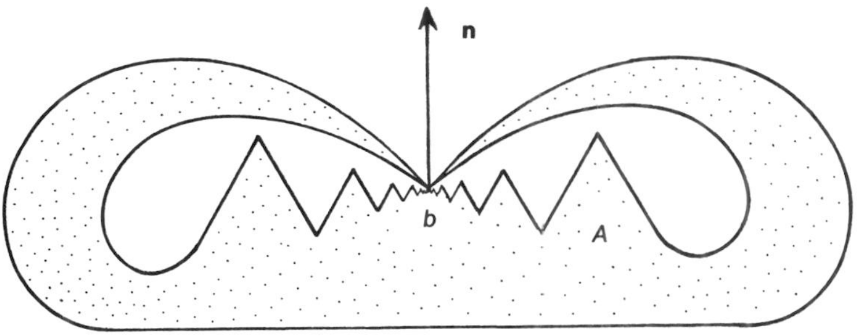

Figure 12.2.1. The generalized normal $\mathbf{n}$ is defined at b because the arms from the sides have density 0 at b.

THEOREM [FEDERER, 4.5.6]. *Let A be an $\mathscr{L}^n$-measurable subset of $\mathbf{R}^n$ such that $T = \partial(\mathbf{E}^n \llcorner A)$ is representable by integration. Then $\|T\| = \mathscr{H}^{n-1} \llcorner \text{domain}\, \mathbf{n}(A, x)$ and, for any Lipschitz vectorfield $\xi(x)$ of compact support,*

$$\int \xi(x) \cdot \mathbf{n}(A, x)\, d\mathscr{H}^{n-1}x = \int_A \operatorname{div} \xi(x)\, d\mathscr{L}^n x.$$

Generalizations to wilder sets have been provided by Harrison.

12.3. Relative Homology [Federer, 4.4]. Suppose $B \subset A$ are C^1, compact submanifolds with boundary of $\mathbf{R}^n$ (or, more generally, compact Lipschitz neighborhood retracts; cf. Federer [1, 4.1.29, 4.4.1, 5.1.6]). Two rectifiable currents S, T in A are *homologous with respect to B* if there is a rectifiable current X in A such that

$$\operatorname{spt}(T - S - \partial X) \subset B.$$

We say that S and T are in the same relative homology class. Given a rectifiable current S, there is a rectifiable current T of least area in its relative homology class.

EXAMPLE 1. Let A be a perturbed solid torus in $\mathbf{R}^3$, let B be its boundary, and let S be a cross-sectional disc. The area minimizer T relatively homologous to S provides a cross-sectional surface of least area. See Figure 12.3.1. The boundary of T is called a *free boundary*.

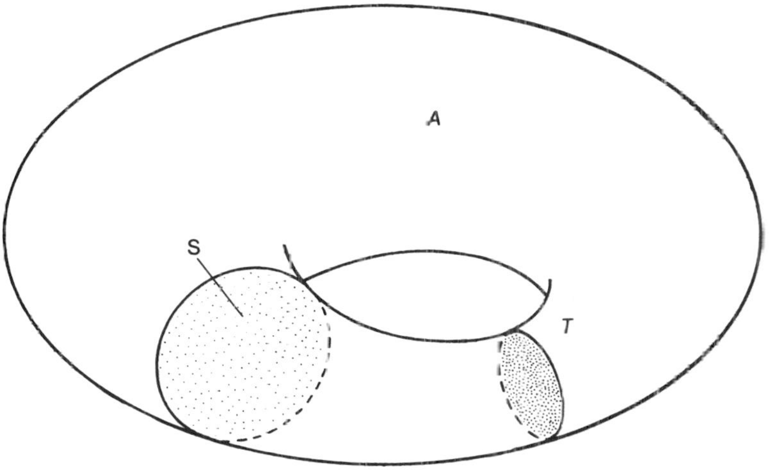

Figure 12.3.1. The area-minimizer T relatively homologous to S provides the least cross-sectional area.

Example 2. Let A be a large, encompassing ball in $\mathbf{R}^3$, let B be the surface of a table (not necessarily flat), and let C be a curve which begins and ends in B. Let S be a surface with $\operatorname{spt}(\partial S - C) \subset B$. The area minimizer T relatively homologous to S provides a surface of least area with the fixed boundary C and additional free boundary in B. It can generally be realized as a soap film. See Figure 12.3.2.

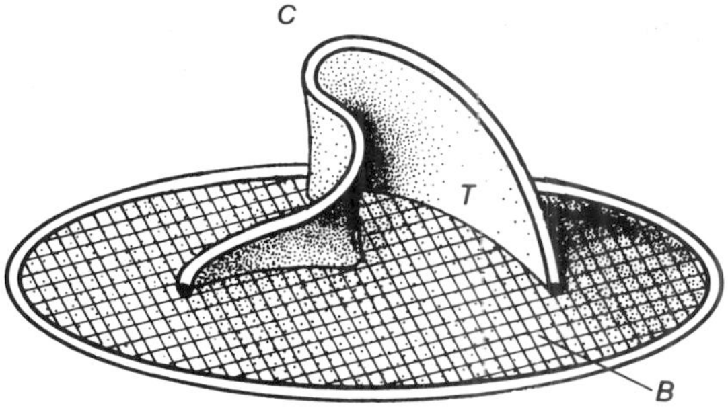

Figure 12.3.2. A soap film minimizing area in its relative homology class.

Proof of Existence. As Brian White pointed out to me, there is a much simpler existence proof than that of Federer [1, 4.4.2, 5.1.6]. Consider a minimizing sequence T_i, viewed as locally integral currents in $M - B$ (cf. 9.1). By compactness, we may assume that the T_i converge to a minimizer T. The hard part is to show that T stays in the same relative homology class. Convergence means that, for some Y and Z_1 in M with small mass,

$$E = T_i - T - Y - \partial Z_1$$

is supported in a small neighborhood of B. Let Y_1 be a minimizer in $\mathbf{R}^N$ with $\partial Y_1 = \partial Y$. Since ∂Y_1 lies in a small neighborhood of B and because $\mathbf{M}(Y_1) \leq \mathbf{M}(Y)$ is small, by monotonicity, 9.5, Y_1 lies in a small neighborhood of B. Let Z_2 be a minimizer in $\mathbf{R}^N$ with $\partial Z_2 = Y - Y_1$. Then $\mathbf{M}(Z_2)$ is small, and hence Z_2 lies in a small neighborhood of M. Let E_1 denote the projection of the cycle $E + Y_1$ onto B; then $E + Y_1 - E_1 = \partial Z_3$, with Z_3 in a small neighborhood of B. We now have

$$T_i - T = E_1 + \partial Z_1 + \partial Z_2 + \partial Z_3,$$

with E_1 in B and the Z_i in a small neighborhood of M. Projecting onto M yields

$$T_i - T = E_1 + \partial Z,$$

with Z in M, so that T lies in the same relative homology class as T_i, as desired.

Isoperimetric Inequality. Very similar arguments provide constants ε and γ, depending on M, such that, if X is an m-dimensional relative cycle in M with $\mathbf{M}(X) \leq \varepsilon$, then there is a rectifiable current Y in M such that $X = \partial Y \pmod B$ and

$$(1) \qquad \mathbf{M}(Y)^{m/(m+1)} + M(\partial Y - X) \leq \gamma \mathbf{M}(X).$$

Remarks on Regularity. Of course, away from B, a relatively homologically area-minimizing rectifiable current enjoys the same regularity as an absolutely area-minimizing rectifiable current. In addition, regularity results are known along the free boundary (see e.g. Taylor [1], Grüter [1, 2], and Hildebrandt). In particular, if A is a smooth Riemannian three-manifold with boundary B, then an unoriented homologically area-minimizing surface T relative to B with $T \subset B$ is a C^1 submanifold with boundary (Taylor [1, 1(7), and Thm. 5], preceded by decomposition into multiplicity-1 boundaries of nested sets as in the Lemma of 10.1 and followed by a maximum principle).

12.4. Functions of Bounded Variation [Federer, 4.5.9; Giusti; Simon 2, §6]. An important class of functions in analysis is the space $\mathrm{BV}^{\mathrm{loc}}$ of functions of locally bounded variation. A real-valued function on $\mathbf{R}^1$ is in $\mathrm{BV}^{\mathrm{loc}}$ if it agrees almost everywhere with a function g of finite total variation on any interval $[a, b]$,

$$\sup\left\{ \sum_{i=1}^{k} |g(x_i) - g(x_{i-1})| : k \in \mathbf{Z},\, a \leq x_0 \leq \cdots \leq x_k \leq b \right\} < \infty,$$

or, equivalently, if the distribution derivative Df is a locally finite measure. Similarly, a real-valued function on $\mathbf{R}^n$ is in $\mathrm{BV}^{\mathrm{loc}}$ if Df is a locally finite (vector-valued) measure. The associated space of currents $\{\mathbf{E}^n \llcorner f\}$ is precisely $\mathbf{N}_n^{\mathrm{loc}} \mathbf{R}^n$, the locally normal currents of codimension 0. Here we give a sampling from Federer's comprehensive theorem on $\mathrm{BV}^{\mathrm{loc}}$.

THEOREM [FEDERER, 4.5.9]. *Suppose* $f \in \mathrm{BV}^{\mathrm{loc}}$.
(13) *If* χ_s *is the characteristic function of* $\{x : f(x) \geq s\}$, *then*

$$Df = \int_{s \in \mathbf{R}} D\chi_s \, ds$$

and

$$|Df| = \int_{s \in \mathbf{R}} |D\chi_s| \, ds$$

almost everywhere.

(31) *If* $n \geq 1$, *then there is a constant* c *such that*

$$\|f - c\|_{L^{n/(n-1)}} \leq n^{-1} \alpha_n^{-1/n} \int |Df|,$$

where α_n *is the volume of the unit ball in* $\mathbf{R}^n$. *If* f *has compact support, then* $c = 0$.

Remarks. The second statement of (13) just says that there is no cancellation in the first. In the notation of geometric measure theory, (13) becomes

$$\partial(\mathbf{E}^n \llcorner f) = \int \partial[\mathbf{E}^n \llcorner \{x: f(x) \geq s\}] \, d\mathscr{L}^1 s$$

and

$$\|\partial(\mathbf{E}^n \llcorner f)\| = \int \|\partial[\mathbf{E}^n \llcorner \{x: f(x) \geq s\}\| \, d\mathscr{L}^1 s.$$

An excellent comprehensive treatment of BV appears in Giusti.

12.5. General Parametric Integrands [Federer 1, 5.1]. In many mathematical and physical problems the cost of a surface depends not only on its area but also on its position or tangent plane direction. The surface energy of a crystal depends on its orientation with respect to the underlying crystal lattice (cf. Taylor [2]). Therefore one considers an integrand $\Phi(x, \xi)$ associating to a rectifiable current T a cost or energy

$$\Phi(T) = \int \Phi\left(x, \vec{T}(x)\right) d\|T\|.$$

The most important case remains the area integrand $A(x, \xi) = |\xi|$. One always requires Φ to be continuous and homogeneous in ξ. Usually Φ is positive (for ξ nonzero), even $(\Phi(x, -\xi) = \Phi(x, \xi))$, and convex $(\Phi(x, \xi_1 + \xi_2) \leq \Phi(x, \xi_1) + \Phi(x, \xi_2))$; i.e., each

$$\Phi_a(\xi) \equiv \Phi(a, \xi)$$

is a norm. (If this norm is given by an inner product, then $\Phi(x, \xi)$ is a Riemannian metric.)

The adjective *parametric* just means that $\Phi(T)$ depends only on the geometric surface T and not on its parameterization (geometric measure theory rarely uses parameterizations at all). Convexity of Φ means each Φ_a is convex,

$$(1) \qquad \Phi_a(\xi_1+\xi_2) \le \Phi_a(\xi_1) + \Phi_a(\xi_2),$$

with the geometric interpretation that the unit ball $\{\Phi_a(\xi) \le 1\}$ is convex. It follows easily that planes are Φ_a minimizing; i.e., if S is a portion of a plane and $\partial R = \partial S$, then

$$(2) \qquad \Phi_a(S) \le \Phi_a(R).$$

When (2) holds, one says F is *semielliptic*.

Uniform convexity of Φ gives an estimate on the strength of the convexity inequality (1),

$$(3) \qquad \Phi_a(\xi_1) + \Phi_a(\xi_2) - \Phi_a(\xi_1+\xi_2) \ge c(|\xi_1| - |\xi_2| - |\xi_1+\xi_2|),$$

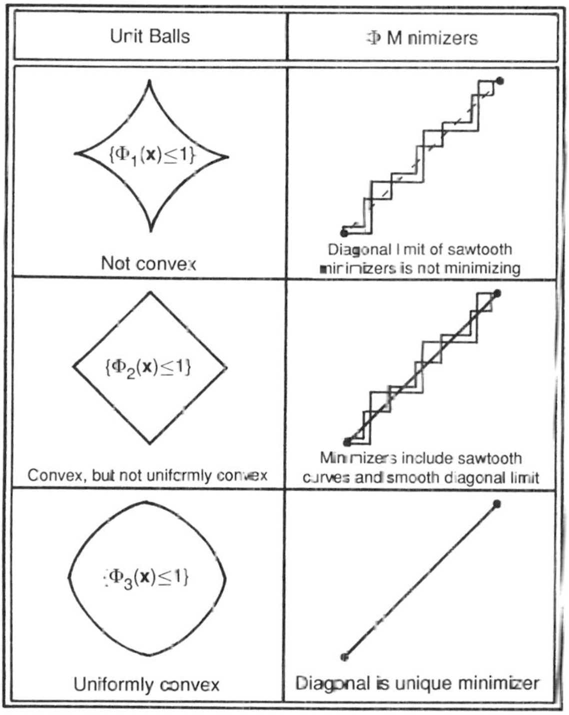

Figure 12.5.1. Three integrands on $\mathbf{R}^2$ for which horizontal and vertical directions are relatively cheap.

with the geometric interpretation for smooth Φ that the unit ball $\{\Phi_a(\xi) \leq 1\}$ has positive inward curvature. It follows easily that, if S is a portion of a plane and $\partial R = \partial S$, then

$$(4) \qquad \Phi_a(R) - \Phi_a(S) \geq \lambda(\mathbf{M}(R) - \mathbf{M}(S)).$$

One says Φ is *elliptic*. Ellipticity, introduced by Almgren, seems to be the right hypothesis for theorems, but much harder to verify directly than uniform convexity. The two notions are equivalent in codimension 1.

All of these notions are invariant under diffeomorphisms of the ambient.

In general, semiellipticity implies lower semicontinuity and the existence of minimizers; ellipticity implies regularity of minimizers (see 8.5). Figure 12.5.1 considers the cost of curves for three integrands $\Phi(\xi)$ on the plane for which horizontal and vertical directions are relatively cheap. Φ_1 is not convex and not lower semicontinuous, so limit arguments can fail to produce minimizers. Φ_2 is borderline convex and admits nonsmooth minimizers. Φ_3 is uniformly convex, and the cheapest path between two points is uniquely a straight line.

CHAPTER 13

Soap Bubble Clusters

Soap bubble clusters illustrate simple mathematical principles. Yet despite notable progress, they defy complete mathematical explanation.

A single soap bubble quickly finds the least-surface-area way to enclose the fixed volume of air trapped inside—the round sphere in Figure 13.0.1. Similarly bubble clusters seek the least-area way to enclose and separate several regions of prescribed volumes. This principle of area minimization alone, implemented on Ken Brakke's Surface Evolver (cf. 13.12), yields computer simulations of bubble clusters, as in Figure 13.0.2, from the video "Computing Soap Films and Crystals" by the Minimal Surface Team at the Geometry Center (formerly the Minnesota Geometry Supercomputer Project).

Do soap bubble clusters always find the absolute least-area shape? Not always. Figure 13.0.3 illustrates two clusters enclosing and separating the same five volumes. In the first, the tiny fifth volume is comfortably nestled deep in the crevice between the largest bubbles. In the second, the tiny fifth volume less comfortably sits between the medium-size bubbles. The first cluster has less surface area than the second. It might be still better to put the smallest bubble around in back.

As a matter of fact, it remains an open question whether the standard double bubble of Figure 13.0.4 is the least-area way to enclose two given volumes, as realized over the course of an undergraduate thesis by Foisy, despite a proof due to White that the solution must be a surface of revolution [Foisy, Thm. 3.4; cf. Morgan 3, Thm. 5.3]. It may seem hard to imagine any other possibilities, until you see J. Sullivan's computer-generated competitor in Figure 13.0.5, which does, however, have more area and is

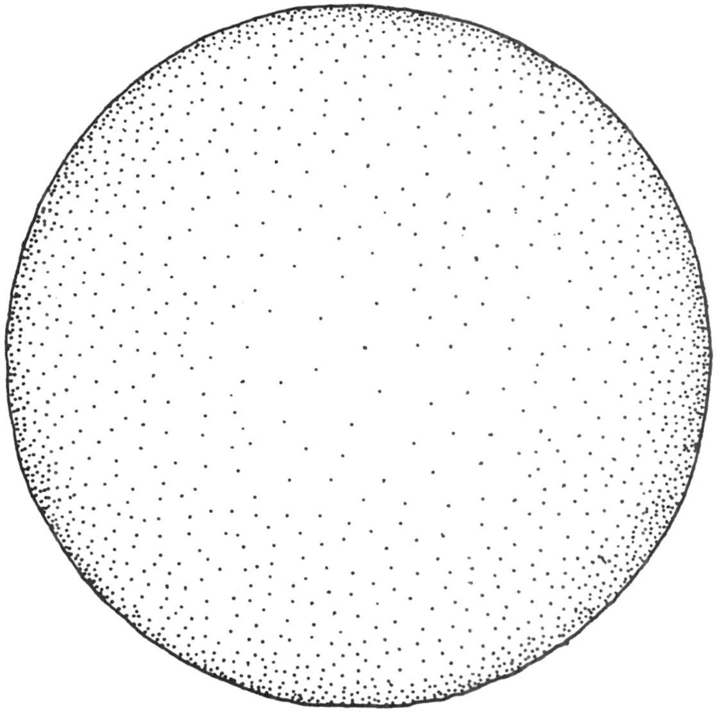

Figure 13.0.1. A spherical soap bubble has found in the least-area way to enclose a given volume of air.

apparently unstable. More generally, both regions might have several components, wrapped around each other.

In general, it is a difficult open question whether each separate region is connected, or whether it might conceivably help to subdivide the regions of prescribed volume, with perhaps half the volume nestled in one crevice here, and the other half in another crevice there.

Similarly, it is an open question whether an area-minimizing cluster may incidentally trap inside "empty chambers," which do not contribute to the prescribed volumes. Figure 13.0.6 shows a 12-bubble with a dodecahedral empty chamber on the inside, obtained by Tyler Jarvis of Mississippi State University using Brakke's Surface Evolver. The computation postulated the empty chamber; without such a restriction, empty chambers probably never occur. Michael Hutchings has proved that in the minimizing double bubble, there are no such empty chambers.

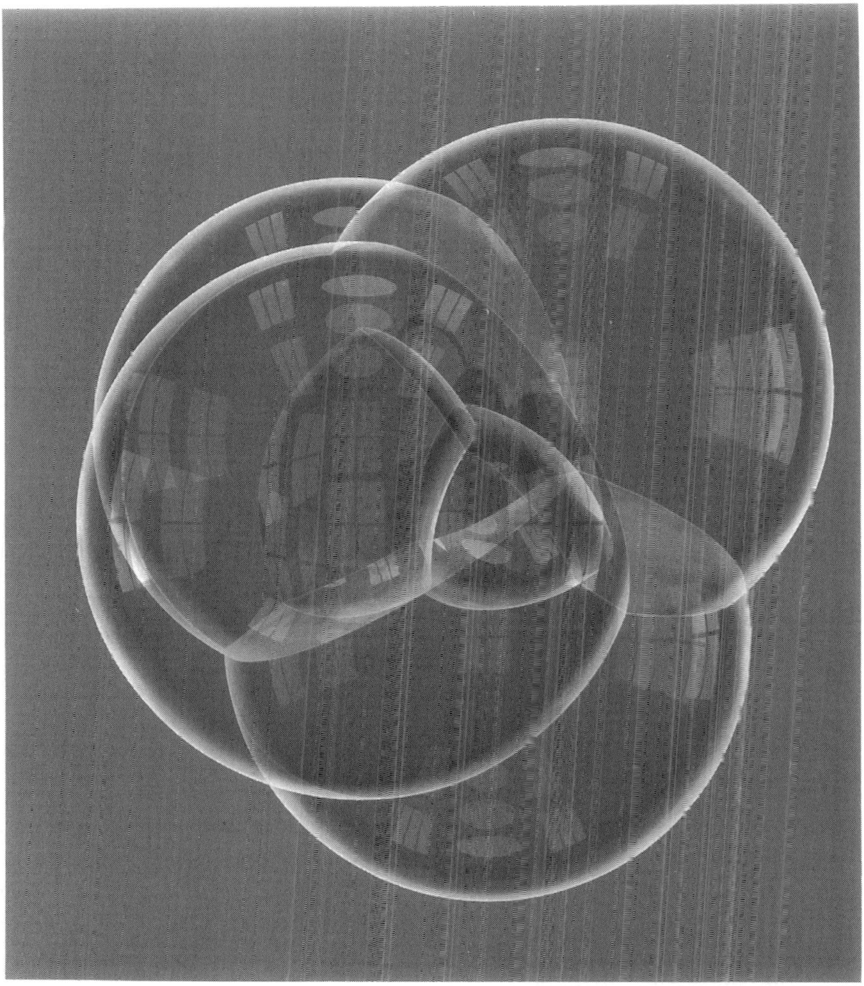

Figure 13.0.2. Bubble clusters seek the least-area way to enclose several volumes of air. Enhanced from "Computing Soap Films and Crystals," a video by the Minimal Surface Team, The Geometry Center. Graphics by John Sullivan.

13.1. Planar Bubble Clusters. Many of the fundamental questions remain open for planar bubble clusters—least-perimeter ways to enclose and separate regions of prescribed areas. A proof of the planar double bubble conjecture appeared in 1993, the work of a group of undergraduates: Joel Foisy, Manuel Alfaro, Jeffrey Brock, Nickelous Hodges, and Jason Zimba (see Figure 13.1.1). (Their work was featured in the 1994 AMS *What's Happening in the Mathematical Sciences*.)

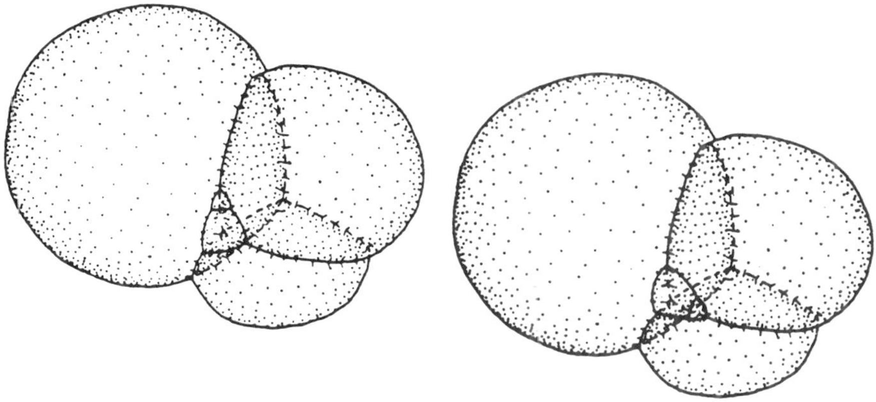

Figure 13.0.3. Soap bubble clusters are sometimes only relative minima for area. These two clusters enclose and separate the same five volumes, but the first has less surface area than the second.

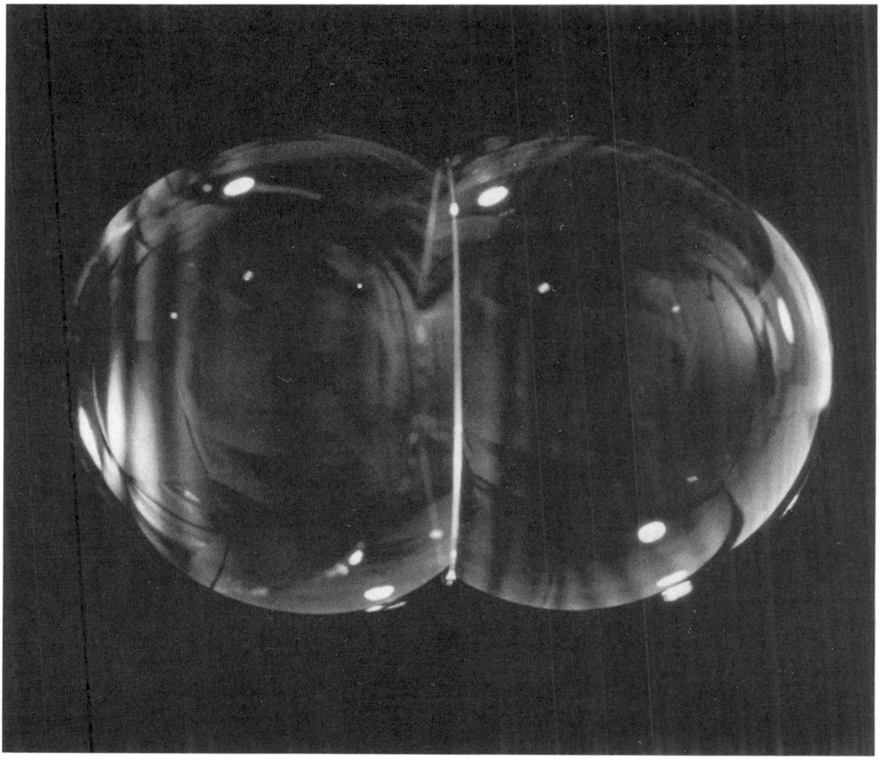

Figure 13.0.4. It remains an open question whether the standard double bubble provides the least-area way to enclose two given volumes. Photo by Jeremy Ackerman, Washington University '96.

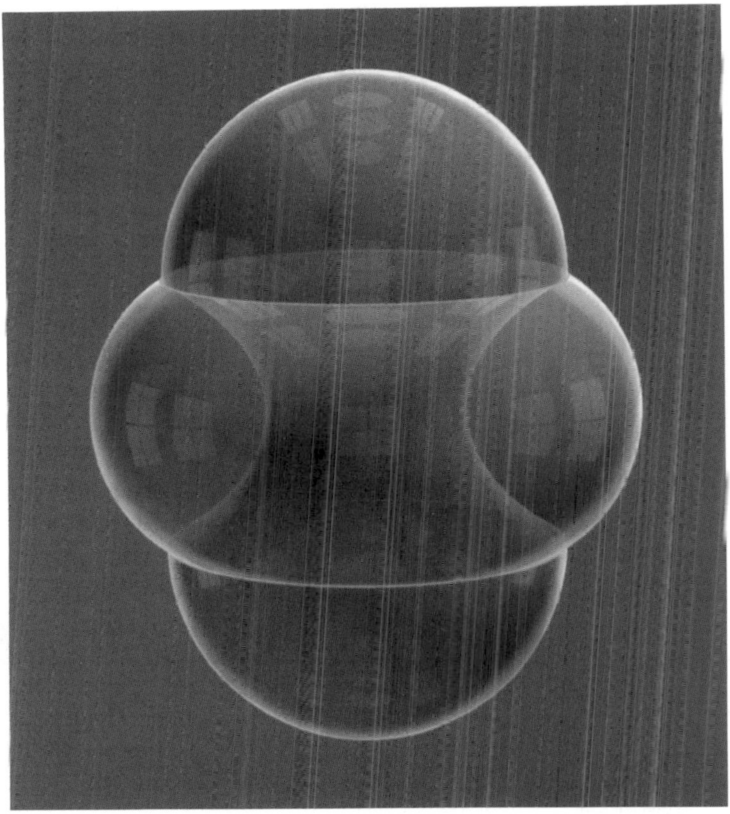

Figure 13.0.5. This nonstandard, computer-generated double bubble has more area and is apparently unstable. Computer graphics by John Sullivan.

In the plane (but not in higher dimensions) the nontrivial general existence theory [Morgan, 14] admits requiring regions to be connected, although they then must be allowed to bump up against each other. For this restricted problem, another group of undergraduates—Christopher Cox, Lisa Harrison, Michael Hutchings, Susan Kim, Janette Light, Andrew Mauer, and Meg Tilton—has obtained the standard triple bubble type in Figure 13.1.2.

13.2. Theory of Single Bubbles. Simplikios's sixth century commentary on Aristotle's *De Caelo* (see Knorr [p. 273], recommended to me by

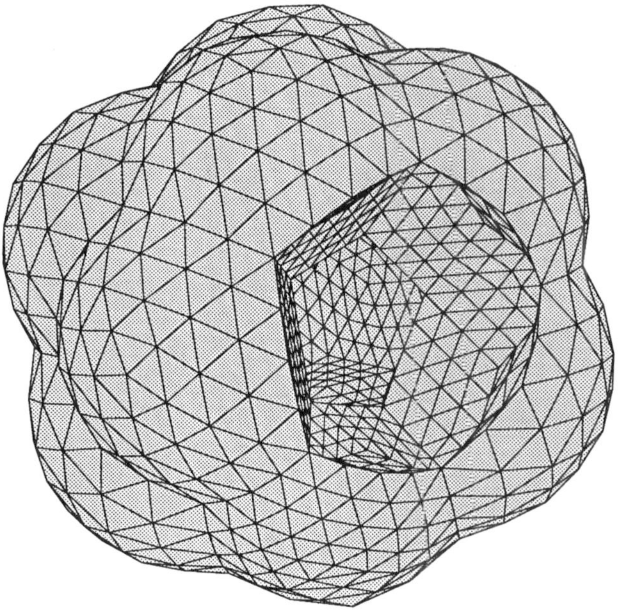

Figure 13.0.6. It is an open question whether area-minimizing clusters may have empty chambers, such as the dodecahedral chamber at the center of this 12-bubble. Graphics by Tyler Jarvis, Mississippi State University.

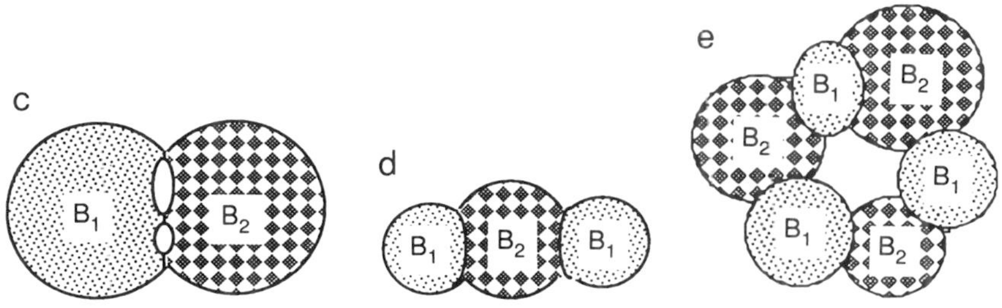

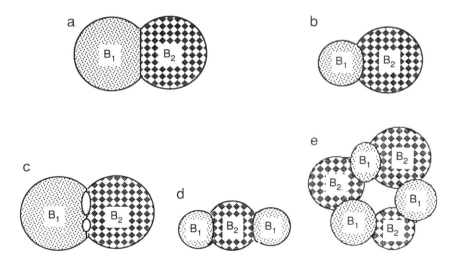

Figure 13.1.1. The standard planar double bubble (a and b) and not some exotic alternative with disconnected regions or empty chambers (c, d, or e) provides the least-perimeter way to enclose and separate two regions of prescribed area, as proved by a group of undergraduates [Foisy *et al.*, 1993].

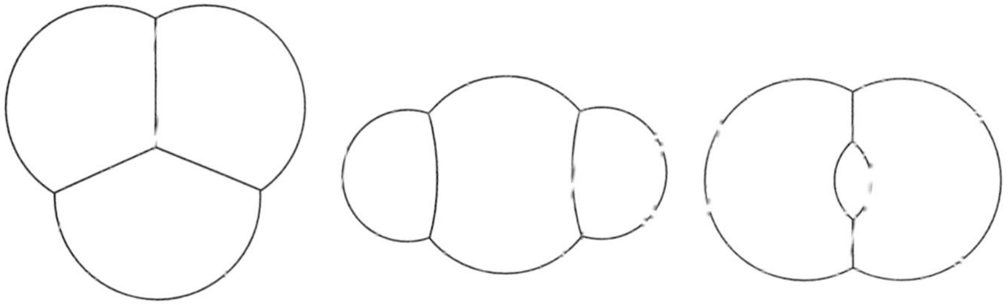

Figure 13.1.2. As the least-perimeter way to enclose and separate *connected* regions of prescribed areas, the solution does not bump up against itself and looks like a rather than b or c, as proved by another group of undergraduates [Cox *et al.*].

D. Fowler), referring incidentally to work no longer extant, reports that

> ...it has been proved...by Archimedes [287–212 B.C.] and Zenodorus [~ 200 B.C.] that of isoperimetric figures the more spacious one...among the solids [is] the sphere.

But Archimedes and Zenodorus considered only a small class of solids, including of course the Platonic solids.

Over 2000 years later H. A. Schwartz (1884) apparently gave the first complete proof by a symmetrization argument. Schwartz symmetrization replaces slices by parallel hyperplanes in $\mathbf{R}^n$ with $(n-1)$-discs centered on an orthogonal axis. De Giorgi [2] (1958) gave a simple completion of an early argument of J. Steiner. Steiner symmetrization replaces slices by parallel lines with intervals centered on an orthogonal hyperplane. Perhaps the simplest known proof is based on the divergence theorem [Gromov, §2.1; Berger, 1, 12.11.4]. For more history, results, and references, see the excellent review by Burago and Zalgaller (especially §10.4). These isoperimetric results generalize to norms more general than area [Gromov, §2.1; Morgan, 11, §10.6; Brothers and Morgan].

Moving beyond bubbles in Euclidean space, Ritoré and Ros have classified the least-area ways to enclose a given volume V in a round $\mathbf{R}P^3$: for small V, a round ball; for large V, its complement; and for middle-sized V, a solid torus centered on an equatorial $\mathbf{R}P^1$.

Although regularity theory (8.5) admits the possibility of singularities of codimension 8 in an area-minimizing single bubble, one might well not expect any. Nevertheless Hsiang (1993) announced an example of a singular bubble in the Cartesian product $\mathbf{H}^7 \times \mathbf{S}^7$ of hyperbolic space with the round sphere. Hsiang uses symmetry to reduce it to a question about curves in the plane. To understand his example, I like to think about the least-perimeter way to enclose a region of prescribed area A on the cylinder $\mathbf{R}^1 \times \mathbf{S}^1$. For small A, the solution is a disc, for large A, the solution is an annular band. Both types occur for a critical value of A, when the minimizer jumps from one type to the other. If the minimizer were continuous in A, it would have to become singular to change type. This is what happens in $\mathbf{H}^7 \times \mathbf{S}^7$.

13.3. Cluster Theory. The existence of soap bubble clusters in $\mathbf{R}^n$ (see Figure 13.3.1) is guaranteed by the following theorem, proved in a more general context by Almgren [1, Theorem V1.2] and specialized and simplified in Morgan [3, §4.4], where details can be found.

Figure 13.3.1. There exists a "soap bubble cluster" providing the least-area way to enclose and separate m regions R_i of prescribed volumes. Photo by Jeremy Ackerman.

A *cluster* consists of disjoint regions $R_1, \dots, R_m$ (n-dimensional locally integral currents of multiplicity 1) with volume $R_i = V_i$, complement R_0, and surface area

$$A = \frac{1}{2} \sum_{i=0}^{m} \mathbf{M}(\partial R_i).$$

(By including R_0, the sum counts each surface twice, before multiplication by $\frac{1}{2}$). A region is not assumed to be connected.

13.4. Existence of Soap Bubble Clusters. *In $\mathbf{R}^n$, given volumes $V_1, \dots, V_m > 0$, there is an area-minimizing cluster of bounded regions R_i of volume V_i.*

To outline a simple proof, we will need a few lemmas. Lemma 13.5, an extremely useful observation of Almgren's (see Almgren [1, VI.2(3)] or Morgan [14, 2.2]), will let us virtually ignore the volume constraints in eliminating wild behavior.

13.5. Lemma. *Given any cluster, there exists $C > 0$, such that arbitrary small volume adjustments may be accomplished inside various small balls at a cost*

$$|\Delta A| \leq C |\Delta V|$$

(*see Figure* 13.5.1).

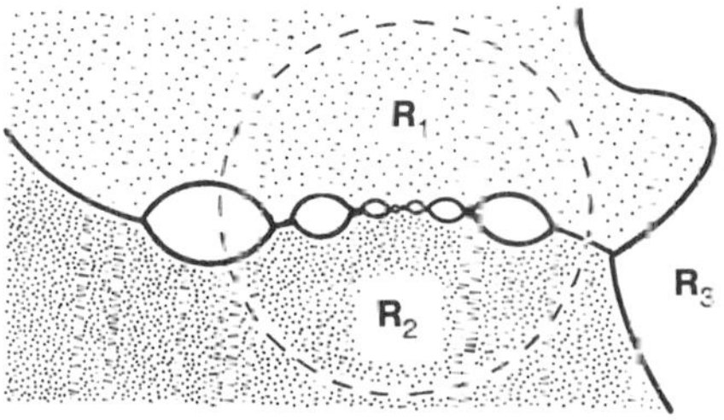

Figure 13.5.1. Gently pushing R_1 into R_2 at a typical border point yields a small volume adjustment with $|\Delta A| \leq C|\Delta V|$.

Remark. There is much freedom in the placement of the balls. Selecting two disjoint sets of such balls now will allow later proofs to adjust volumes at such locations, away from the main argument.

Proof sketch. By the Gauss–Green–De Giorgi–Federer Theorem, 12.2, at almost all points of ∂R_i, R_i has a measure-theoretic exterior normal and the approximate tangent cone is a half-space. Hence at almost every point of ∂R_1, for example, this half-space fits up against that of some other R_i, say R_2. Gently pushing R_1 into R_2 costs $|\Delta A| \le C_{12}|\Delta V|$. Combining over many such neighboring pairs yields an arbitrary small volume adjustment with $|\Delta A| \le C|\Delta V|$.

13.6. Lemma. *An area-minimizing cluster is bounded in* $\mathbf{R}^n$.

Proof. Let $V(r)$ denote the volume outside $\mathbf{B}(0,r)$; let $A(r)$ denote the area outside $\mathbf{B}(0,r)$. Truncation at almost any radius r saves $A(r)$, requires patching by the slice $\langle R_i, u, r+\rangle$ with $u(x) = |x|$ and

$$\mathbf{M}\langle R_i, u, r+\rangle \le |V'(r)|$$

(4.11(3)), and requires replacing lost volume at a cost of $CV(r)$ by Lemma 13.5. Therefore

$$|V'(r)| + CV(r) \ge A(r). \tag{1}$$

On the other hand, application of the isoperimetric inequality, 5.3, to the exterior of $\mathbf{B}(0,r)$ yields

$$|V'(r)| + A(r) \ge \gamma V(r)^{(n-1)/n}. \tag{2}$$

Adding inequalities (1) and (2) yields

$$2|V'(r)| \ge -CV(r) + \gamma V(r)^{(n-1)/n} \ge \tfrac{1}{2}\gamma V(r)^{(n-1)/n}$$

for large r. If $V(r)$ is never 0,

$$n\left(V^{1/n}\right)' = V^{-(n-1)/n} V' \le -c < 0$$

for almost all large r, which contradicts positive and nonincreasing V.

13.7. Sketch of Proof of Theorem 13.4. The main difficulty is that volume can disappear to infinity in the limit. First we show that we can preserve some fraction of the volume. Let $\mathscr{C}_\alpha$ be a minimizing sequence of clusters with the prescribed volumes V_i. We claim there are constants $S, \delta > 0$, such that if $\mathbf{R}^n$ is partitioned into cubes K_i of edge length S, then for some K_i,

$$\operatorname{vol}(R_{1,\gamma} \llcorner K_i) \ge \delta V_1. \tag{1}$$

Indeed choose S large enough so that if $\operatorname{vol}(R \llcorner K_i) \le V_1$, then

$$\operatorname{area}(\partial R \llcorner K_i) \ge \gamma\left(\operatorname{vol}(R \llcorner K_i)\right)^{(n-1)/n},$$

for some isoperimetric constant γ (12.3(1)). Let $\delta \geq \gamma / A$. Then for each K_i,

$$\operatorname{area}(\partial R_{1,\alpha} \llcorner K_i) \geq \frac{\gamma}{\max_i (R_{1,\alpha} \llcorner K_i)} \operatorname{vol}(R_{1,\alpha} \llcorner K_i).$$

Summing over i yields

$$A \geq \frac{\gamma}{\max_i (R_{1,\alpha} \llcorner K_i)} V_1,$$

$$\max_i (R_{1,\alpha} \llcorner K) \geq \gamma V_1 / A \geq \delta V_1,$$

proving the claim. Therefore, by translating the $\mathscr{C}_\alpha$, we may assume that for some fixed $r > 0$,

(2) $$\operatorname{vol}(R_{1,\alpha} \llcorner \mathbf{B}(0, r)) \geq \delta V_1.$$

By a compactness argument (see 9.1), we now may assume that the $\mathscr{C}_\alpha$ converge to a limit cluster $\mathscr{C}$. By (2),

(3) $$\operatorname{vol}(R_1) \geq \delta V_1.$$

The second step is to show that $\mathscr{C}$ is area minimizing for its volume. The argument involves restricting the $\mathscr{C}_\alpha$ to an increasing sequence of balls (discarding everything in their exteriors), judiciously chosen to make the required patching inconsequential. Lemma 13.5 is required to preserve the volume constraints. By Lemma 13.6, $\mathscr{C}$ is bounded.

Now if there was no volume loss to infinity, $\mathscr{C}$ solves our problem. If there was a volume loss, repeat the whole process with translations of the discarded material. Since each repetition recovers a fixed fraction of missing volume, countably many repetitions can capture the total volume and yield a solution. (Since each is bounded, they do fit in $\mathbf{R}^n$). Since the conglomerate solution must be bounded by Lemma 13.6, we conclude that only finitely many repetitions were actually needed.

General Ambient Manifolds. Theorem 13.4 holds in any smooth Riemannian manifold M with compact quotient M/Γ by the isometry group Γ.

Regularity results for minimizing clusters begin with the following.

13.8. Proposition. *In a minimizing cluster, the rectifiable set* $S = \bigcup(\partial R_i)$ *is* $(\mathbf{M}, \varepsilon, \delta)$ *minimal*:

$$\mathscr{H}^{n-1}(S) \leq (1 + \varepsilon(r)) \mathscr{H}^{n-1}(\varphi(S)),$$

where $\varepsilon(r) = 3Cr$, $r \leq \delta$.

Proof. The proof depends on Lemma 13.5, which provides volume adjustments at cost $|\Delta A| \le C|\Delta V|$. δ must be chosen small enough so that a δ-ball has small volume in the sense of Lemma 13.5 and so that any δ-ball is disjoint from a set of balls used to readjust volumes.

Consider a Lipschitz deformation inside an r-ball with $r \le \delta$. The total volume distortion in moving S to $\varphi(S)$ is at most $r(\mathscr{H}^{n-1}(S) + \mathscr{H}^{n-1}(\varphi(S)))$ and certainly less than the volume of the r-ball. By Lemma 13.5, the volumes may be readjusted elsewhere at cost

$$|\Delta A| \le C|\Delta V| \le Cr\big(\mathscr{H}^{n-1}(S) + \mathscr{H}^{n-1}(\varphi(S))\big).$$

Figure 13.8.1. Jean Taylor and Fred Almgren at their wedding, with Rob and Ann Almgren. Photograph courtesy of the Almgren-Taylor family.

Since the original cluster is minimizing,

$$\mathscr{H}^{n-1}(S) \le \mathscr{H}^{n-1}(\varphi(S)) + Cr(\mathscr{H}^{n-1}(S) - \mathscr{H}^{n-1}(\varphi(S)),$$

$$\mathscr{H}^{n-1}(S) \le \frac{1+Cr}{1-Cr}\mathscr{H}^{n-1}(\varphi(S)) \le (1+3Cr)\mathscr{H}^{n-1}(\varphi(S))$$

for $r \le \delta$ small. Therefore S is $(\mathbf{M}, \varepsilon, \delta)$ minimal for $\varepsilon(r) = 3Cr$.

In 1976 Jean Taylor gave a definitive mathematical explanation of the structure of soap bubble clusters recorded more than a century before by J. A. F. Plateau, based on Fred Almgren's theory of $(\mathbf{M}, \varepsilon, \delta)$-minimal sets. A beautiful description of Taylor's work appears in a *Scientific American* article by Taylor and Almgren, who supervised her Ph.D. thesis. The latest account is by Kanigel. As another nice result, Almgren and Taylor were married (see Figure 13.8.1).

13.9. Regularity of Soap Bubble Clusters in R^3 [Taylor, 4]. *A soap bubble cluster $\mathscr{C}$ in $\mathbf{R}^3$ ($(\mathbf{M}, \varepsilon, \delta)$-minimal set) consists of real analytic constant-mean-curvature surfaces meeting smoothly in threes at 120° angles along smooth curves, in turn meeting in fours at angles of* $\cos^{-1}(-1/3) \approx 109°$.

Remark. The singular curves were proved $C^{1,\alpha}$ by Taylor [4]; C^∞ by Nitsche [1]; and real-analytic by Kinderlehrer, Nirenberg, and Spruck [Thm. 5.1].

Comments on Proof. Consider a linear approximation or tangent cone C at any singularity (which exists by monotonicity and further substantial arguments). By scaling, C is $(\mathbf{M}, \varepsilon, \delta)$ minimal for $\varepsilon = 0$ and $\delta = \infty$. C must intersect the unit sphere in a "net" of geodesic curves meeting in threes at 120°, an extension of the more familiar fact that shortest networks meet only in threes at 120°. (This angle comes from a balancing condition for equilibrium. A junction of four curves could be profitably deformed to two junctions of three as in Figure 11.3.1). In 1964 Heppes, apparently unaware of much earlier incomplete work of Lamarle (1864), found all ten such geodesic nets, pictured in Figure 13.9.1. Taylor, adding a final case to complete the work of Lamarle, showed all but the first three cones to be unstable by exhibiting area-decreasing deformations as in Figure 13.9.2. In an ironic twist of fate, in the fourth case, the comparison surface provided by Taylor has more area than the cone; the correct comparison surface, which reduces area by pinching out a flat triangular surface in the center, had been given correctly by Lamarle.

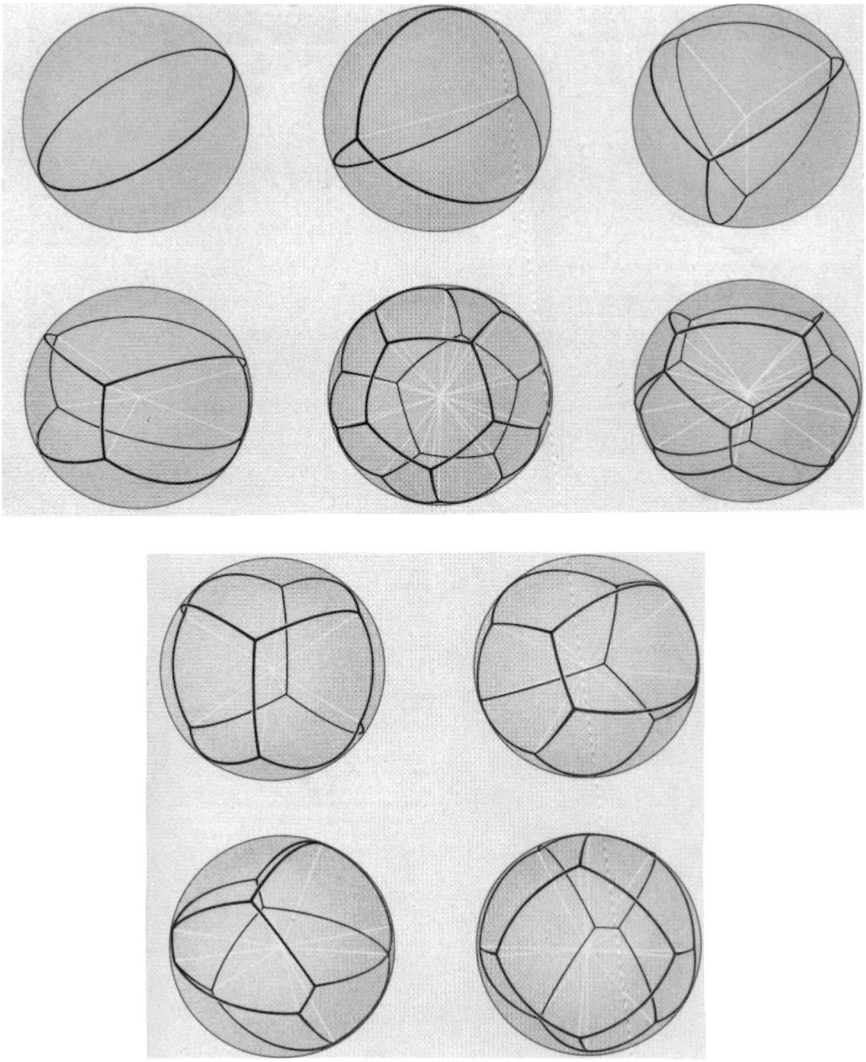

Figure 13.9.1. On the sphere there are exactly 10 nets of geodesics meeting in threes at 120°, providing 10 candidate cone models for soap bubble structures. Reprinted with permission from Almgren and Taylor.

Thus, for the approximating cone, there are just three possibilities, corresponding to a smooth surface or the two asserted types of singularities.

The really hard part is to show that the cone is a good enough approximation to the original soap film. Taylor uses a deep method of Reifenberg

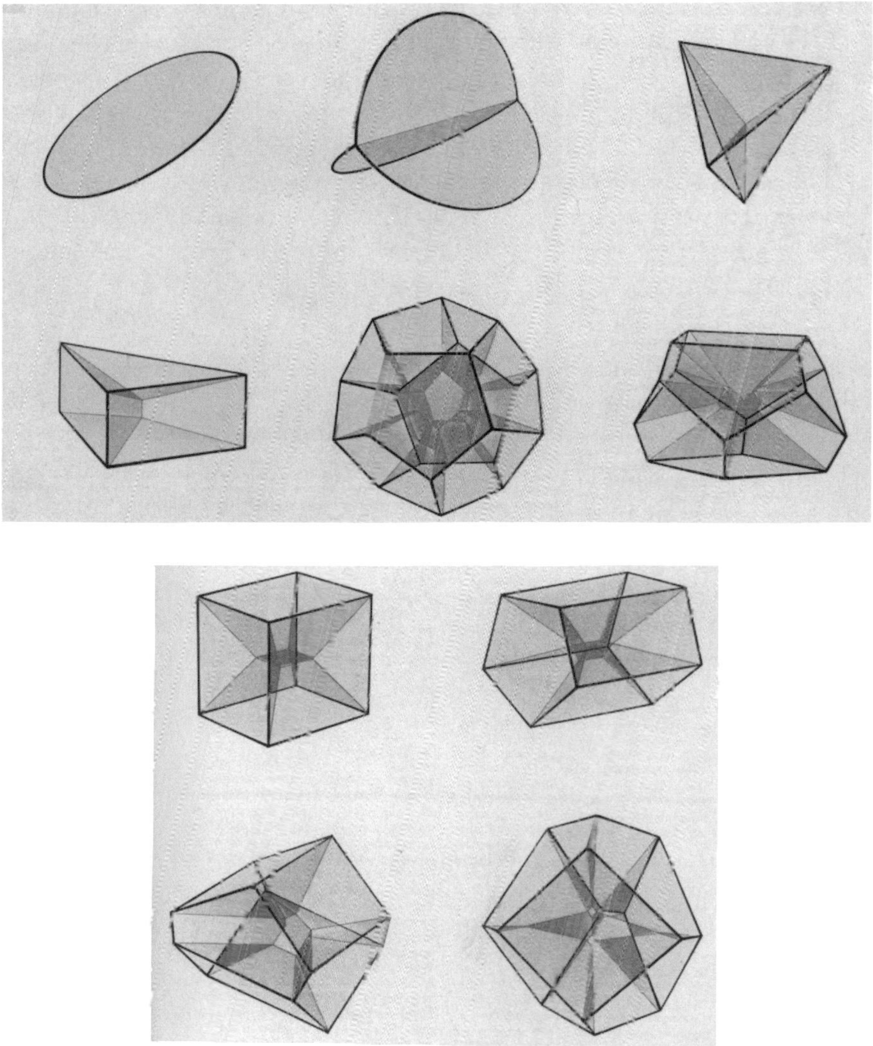

Figure 13.9.2. All but the first three cones are unstable, as demonstrated by the pictured deformations of less area. The fourth one actually should have a horizontal triangle instead of a vertical line segment in the center. Copyright © 1976 by Scientific American, Inc. All rights reserved.

[1–3], requiring the verification of a certain "epiperimetric inequality," which says roughly that cones near the special three are not too close to being minimizers themselves.

Of course, where the surface is regular, a classical variational argument yields constant mean curvature and hence real analyticity.

13.10. Cluster Regularity in Higher Dimensions. Almgren [1, Theorem III.3(7)] proved that soap bubble clusters ($(\mathbf{M}, \varepsilon, \delta)$-minimal sets) in $\mathbf{R}^n$ ($n \geq 3$) are $C^{1,\alpha}$ almost everywhere. For refinements see [White, 6]. Brakke [2] has classified the polyhedral $(\mathbf{M}, 0, \infty)$-minimal cones in $\mathbf{R}^4$, which include the cone over the hypercube. It is conjectured that there are no nonpolyhedral $(\mathbf{M}, 0, \infty)$-minimal cones below $\mathbf{R}^8$. Almgren's memoir does not treat the case of $\mathbf{R}^2$ [Almgren, 1, IV.3 (1), p. 96; Morgan, 7], which is different in many ways. Only in $\mathbf{R}^2$ does existence theory provide the option of requiring that regions be connected [Morgan, 14].

13.11. Minimizing Surface and Curve Energies. In some materials, not only interfacial surfaces but also singular curves may carry energetic costs. This modification alters behavior qualitatively as well as quantitatively, with four surfaces meeting along a singular curve, for example. Some results on existence, regularity, and structure appear in Morgan and Taylor and in Morgan [17, 3].

13.12. Closing Remarks. This chapter began with my AMS–MAA address in San Francisco, 1991, available on video [Morgan, 4] and written up in Morgan [9, 8].

Mathematics has gone beyond minimizers in equilibrium to dynamical processes such as crystal growth, with impressive progress in theory, computation, and graphics. See, for example, the survey on *Geometric Models of Crystal Growth*, by Taylor, Cahn, and Handwerker, or Brakke's *The Surface Evolver*. (A package containing the source code, the manual, and sample data files is freely available by anonymous ftp in the file pub/software/evolver/evolver.tar.z on the machine *geom.umn.edu*.)

13.13. Kelvin disproved by Weaire and Phelan. 1994 brought striking news of the disproof of Lord Kelvin's 100-year-old conjecture by Denis Weaire and Robert Phelan of Trinity College, Dublin. Kelvin sought the least-area way to partition all of space into regions of unit volumes. (Since the total area is infinite, least area is interpreted to mean that there is no area-reducing alteration of compact support preserving the unit volumes.) His basic building block was a truncated octahedron, with its six square faces of truncation and eight remaining hexagonal faces, which packs perfectly to fill space as suggested by Figure 13.13.1. (The regular dodecahedron, with its twelve pentagonal faces, has less area, but it does not pack.) The whole structure relaxes slightly into a curvy equilibrium, which is Kelvin's candidate. All regions are congruent.

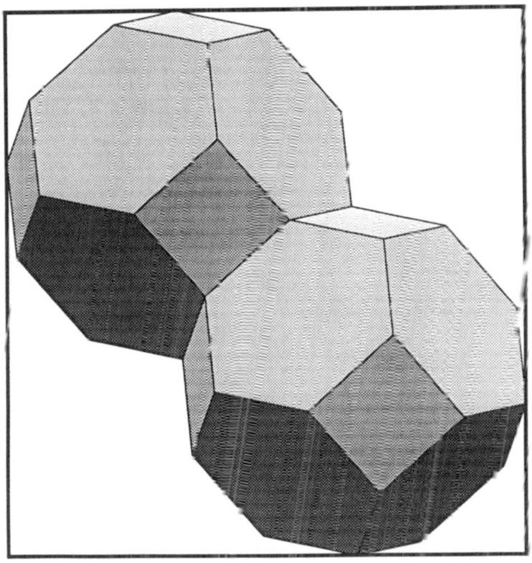

Figure 13.13.1. Lord Kelvin conjectured that the least-area way to partition space into unit volumes uses relaxed truncated octahedra. Computer graphics by Ken Brakke [5] of the Geometry Center.

Weaire and Phelan recruited two different building blocks from certain chemical clathrate compounds: the regular dodecahedron and a tetrakaidecahedron with 12 pentagonal faces and 2 hexagonal faces. The tetrakaidecahedra are arranged in three orthogonal stacks, stacked along the hexagonal faces, as in Figure 13.13.2. The remaining holes are filled by

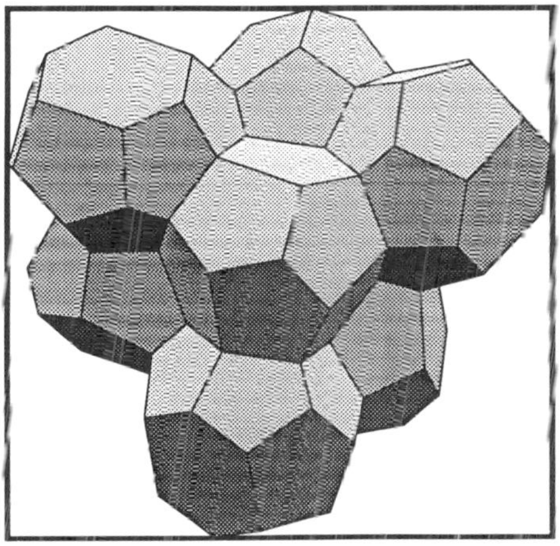

Figure 13.13.2. The relaxed stacked tetrakaidecahedra and occasional dodecahedra of Weaire and Phelan beat Kelvin's conjecture by about 0.3%. Computer graphics by Ken Brakke [5] of the Geometry Center.

dodecahedra. Again, the structure is allowed to relax into a stable equilibrium. Computation on the Brakke Evolver shows an improvement over Kelvin's conjecture of about 0.3%. Weaire and Phelan thus provide a new conjectured minimizer. Weaire's popular account in *New Scientist* gives further pictures and details.

In greater detail, the centers of the polyhedra are at the points of a lattice with the following coordinates modulo 2:

0	0	0
1	1	1
.5	0	1
1.5	0	1
0	1	.5
0	1	1.5
1	.5	0
1	1.5	0

Given a center, the corresponding polyhedral region is just the "Voronoi cell" of all points closer to the given center than to any other center. The relaxation process also needs to slightly adjust the volumes to make them all 1.

Incidentally, Kelvin's particular truncated octahedron is actually a scaled "permutohedron," the convex hull of the 24 permutations of (1, 2, 3, 4) in $\mathbf{R}^3 = \{\mathbf{x} \in \mathbf{R}^4 \colon \Sigma x_i = 10\}$.

Proving the new Weaire–Phelan conjecture could take a while. After all, the single bubble was proved minimizing by Schwartz in 1884, and the double bubble remains conjectural in 1994. Will Weaire–Phelan's "infinite bubble" take another century?

Solutions to Exercises

Chapter 2

2.1.

$$\mathscr{I}^1(I) = \frac{1}{\beta(2,1)} \int_{p\in \mathbf{O}^*(2,1)} \int_{y\in \operatorname{Im} p} N(p|I, y)\, d\mathscr{L}^1 y\, dp$$

$$= \frac{\pi}{2}\int_0^{\pi} |\cos\theta| \frac{d\theta}{\pi} = 1.$$

2.2. Coverings by n intervals of length $1/n$ show that $\mathscr{H}^1(I) \le 1$. Suppose $\mathscr{H}^1(I) < 1$. Then there is a covering $\{S_j\}$ of I with

$$\sum \operatorname{diam} S_j < 1.$$

By slightly increasing each diam S_j if necessary, we may assume that the S_j are open intervals (a_j, b_j). Since I is compact, we may assume that there are only finitely many. We may assume that none contains another. Finally we may assume that $a_1 < a_2 < \cdots < a_n$ and hence $b_j > a_{j+1}$. Now

$$\sum_{j=1}^{n} \operatorname{diam} S_j = \sum_{j=1}^{n} (b_j - a_j) \ge \sum_{j=1}^{n-1} (a_{j+1} - a_j) + (b_n - a_n)$$

$$= b_n - a_1 > 1,$$

the desired contradiction.

2.3. Covering $[-1,1]^n$ by $(2N)^n$ cubes of side $1/N$ and radius $\sqrt{n}/2N$ yields

$$\mathscr{H}^n(\mathbf{B}^n(0,1)) \le \mathscr{H}^n([-1,1]^n) \le \lim (2N)^n \alpha_n (\sqrt{n}/2N)^n = \alpha_n n^{n/2} < \infty.$$

2.4. For each $\delta > 0$, there is a cover $\{S_j(\delta)\}$ of A with $\operatorname{diam}(S_j(\delta)) \leq \delta$ and

$$\sum \alpha_m \left(\frac{\operatorname{diam} S_j(\delta)}{2} \right)^m \leq \mathscr{H}^m(A) + \varepsilon < \infty$$

Consequently,

$$\begin{aligned} \lim \sum \alpha_k \left(\frac{\operatorname{diam} S_j(\delta)}{2} \right)^k &= \frac{\alpha_k}{\alpha_m} \lim \sum \alpha_m \left(\frac{\operatorname{diam} S_j(\delta)}{2} \right)^m \left(\frac{\operatorname{diam} S_j(\delta)}{2} \right)^{k-m} \\ &\leq \frac{\alpha_k}{\alpha_m} (\mathscr{H}^m(A) + \varepsilon) \lim \left(\frac{\delta}{2} \right)^{k-m} = 0. \end{aligned}$$

Therefore $\mathscr{H}^k(A) = 0$.

It follows that for a fixed set A, there is a nonnegative number d such that

$$\mathscr{H}^m(A) = \begin{cases} \infty & \text{if } 0 \leq m < d, \\ 0 & \text{if } d < m < \infty. \end{cases}$$

All four definitions of the Hausdorff dimension of A yield d. Incidentally, $\mathscr{H}^d(A)$ could be anything: 0, ∞, or any positive real number, depending on what A is.

2.5. The 3^j triangular regions of side 3^{-j} which make up A_j provide a covering of A with $\Sigma \alpha_1 (\operatorname{diam}/2)^1 = 1$. Hence $\mathscr{H}^1(A) \leq 1$.

The opposite inequality is usually difficult, but here there is an easy way. Let Π denote projection onto the x-axis. Then $\mathscr{H}^1(A) \geq \mathscr{H}^1(\Pi(A)) = 1$.

2.6. (a) C can be covered by 2^n intervals of length 3^{-n}.

(b) Suppose $\mathscr{H}^m(C) < \alpha_m / 2^m$, so there is a covering by intervals S_i with $\Sigma(\operatorname{diam} S_i)^m < 1$. By slight enlargement, we may assume that each S_i is open. Since C is compact, we may assume that $\{S_i\}$ is finite, of minimal cardinality. If no S_i meet both halves of C, a scaling up of the covering of one of the two halves yields a cheap covering of C of smaller cardinality, a contradiction. If exactly $p \geq 1$ S_i's meet both halves of C, replace each by

$$\left\{ x \in S_i \colon x \leq \tfrac{1}{3} \right\} \quad \text{and} \quad \left\{ x \in S_i \colon x \geq \tfrac{2}{3} \right\}.$$

For $0 \leq s,\ t \leq \frac{1}{3}$,

$$f(s,t) \equiv \left(s + \tfrac{1}{3} + t \right)^m - s^m - t^m \geq 0,$$

because the partial derivatives are negative and $f(\frac{1}{3}, \frac{1}{3}) = 0$; therefore, the new covering still satisfies $\Sigma(\operatorname{diam} S_i)^m < 1$. $p \leq 2$ S_i's meet both halves, since any in addition to the leftmost and rightmost such would be superfluous. If the original covering had (minimal) cardinality n, the new one has cardinality at most $n+p$. A scaling up of the covering of one of the two halves yields a cheap covering of C of cardinality at most $(n+p)/2$. Since $(n+p)/2 \geq n$, $n \leq p$. If $n=1$, $\Sigma(\text{diam } S_i)^m \geq 1$, a contradiction. Otherwise $n=p=2$, each S_i has a diameter of at least $\frac{1}{3}$, and $\Sigma(\text{diam } S_i)^m \geq 1$, a contradiction.

2.7.

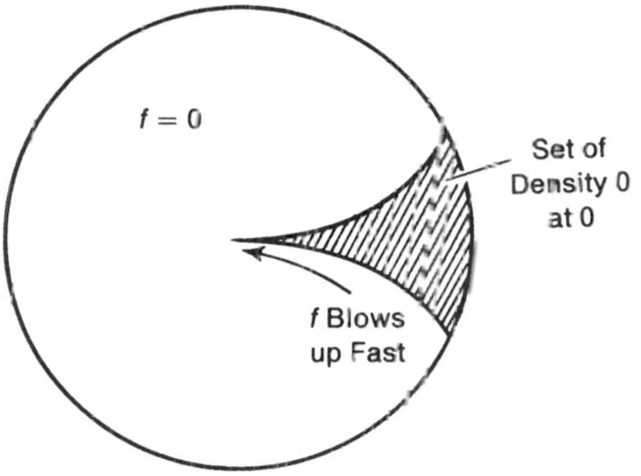

2.8. Let $\varepsilon > 0$. Suppose $\mathbf{0}$ is a Lesbesgue point of f. Then

$$\frac{1}{\alpha_m r^m}\int_{\mathbf{B}^m(\mathbf{0},r)} |f(x)-f(\mathbf{0})|\, d\mathscr{L}^m \to 0.$$

Consequently,

$$\frac{1}{\alpha_m r^m}\mathscr{L}^m\{x\in \mathbf{B}^m(\mathbf{0},r): |f(x)-f(\mathbf{0})| \geq \varepsilon\} \to 0.$$

Therefore, f is approximately continuous at $\mathbf{0}$.

2.9. Following the hint, let $a \in \bigcap E_i$. It suffices to show that f is approximately continuous at a, since by Corollary 2.9 almost every point lies in $\bigcap E_i$.

Given $\varepsilon > 0$, choose

$$f(a)-\varepsilon < q_i < f(a) < q_j < f(a)+\varepsilon.$$

Then

$$\Theta(\{|f(x)-f(a)| \geq \varepsilon\},a) \leq \Theta(\{f(x) < q_i\},a) + \Theta(\{f(x) > q_j\},a) = 0+0,$$

because $a \in E_i$ and $a \in E_j$.

Chapter 3

3.1. Let $\{q_i\}$ be an enumeration of the rationals and let

$$f(x) = \sum_{i=1}^{\infty} 2^{-i}|x - q_i|.$$

3.2. Given $\varepsilon > 0$, f is approximately differentiable at the points of density 1 of $\{x \in A\colon f(x) = g(x)\}$, i.e., everywhere except for a set of measure $< \varepsilon$.

3.3. On all of $\mathbf{R}$, one can just take $f(x) = x^2$. On $[-1,1]$, one can take $f(x) = \sqrt[3]{x}$.

3.4.

$$J_1 f = \begin{cases} 1 & \text{for } r \geq 1 \\ 1/r & \text{for } r \leq 1 \end{cases}$$

3.5.

$$\begin{aligned} J_1 f &\equiv \max\{ Df(u)\colon u \text{ unit vector}\} \\ &= \max\{\nabla f \cdot u\} = |\nabla f|. \end{aligned}$$

3.6.

$$\mathscr{H}^2\left(\mathbf{S}^2(\mathbf{0},1)\right) = \int_0^{2\pi}\int_0^{\pi} J_2 f \, d\varphi \, d\theta.$$

$$J_2 f = \sin\varphi.$$

The answer is 4π.

3.7.

$$\begin{aligned} \text{LHS} &= \int_A J_1 f \, d\mathscr{L}^3 = \int_{\mathbf{B}(\mathbf{0},R)} 2r \, d\mathscr{L}^3 = \int_0^R (2r)\left(4\pi r^2\right) dr \\ &= 2\pi R^4. \end{aligned}$$

$$\text{RHS} = \int_0^{R^2} 4\pi y \, dy = 2\pi R^4.$$

3.8. (a) Apply the area formula, 3.13, to $f\colon E \times [0,1] \to C$ given by $f(x,t) = xt$.

$$\mathscr{H}^{m+1}(C) = \int_{E\times[0,1]} J_{m+1} f = \int_{E\times[0,1]} t^m = \frac{a_0}{m+1}.$$

Alternatively, apply the coarea formula, 3.13, to $f\colon C \to \mathbf{R}$ given by $f(x) = |x|$. Then

$$\mathscr{H}^{m+1}(C) = \int_0^1 a_0 y^m = \frac{a_0}{m+1}.$$

(b)

$$\begin{aligned} \Theta^{m+1}(C,\mathbf{0}) &= \lim \frac{\mathscr{H}^{m+1}(C \cap B(\mathbf{0},r))}{\alpha_{m+1} r^{m+1}} \\ &= \lim \frac{a_0 r^{m+1}/(m+1)}{\alpha_{m+1} r^{m+1}} \\ &= \frac{a_0}{\alpha_{m+1}(m+1)}. \end{aligned}$$

(c) The cone over the closure of $\{x \in E: \Theta^m(E,x) \neq 0\}$ equals the cone over $\{x \in E: \mathscr{H}^m(B(x,r) \cap E) > 0 \text{ for all } r > 0\}$.

3.9. Let $\{q_i\}$ be an enumeration of the points in $\mathbf{R}^3$ with rational coordinates. Let $E = \bigcup_{i=1}^{\infty} \mathbf{S}(q_i, 2^{-i})$. By 3.12, $\Theta^2(E,x) = 1$ for almost all $x \in E$. It follows that $\{x \in \mathbf{R}^3: \Theta^2(E,x) = 1\}$ is dense in $\mathbf{R}^3$.

Chapter 4

4.1. $-6\mathbf{e}_{134} - 12\mathbf{e}_{234}$.

4.2. One possibility is

$$u = (-1, 0, 1, -1),$$

$$v = (0, -1, 1, -1),$$

$$w = \left(-\frac{1}{\sqrt{3}}, 0, \frac{1}{\sqrt{3}}, -\frac{1}{\sqrt{3}}\right).$$

$$z = \left(\frac{2}{\sqrt{15}}, -\frac{3}{\sqrt{15}}, \frac{1}{\sqrt{15}}, -\frac{1}{\sqrt{15}}\right).$$

$$u \wedge v = \mathbf{e}_{12} - \mathbf{e}_{13} + \mathbf{e}_{14} + \mathbf{e}_{23} - \mathbf{e}_{24}$$

$$w \wedge z = \frac{1}{\sqrt{5}}\mathbf{e}_{12} - \frac{1}{\sqrt{5}}\mathbf{e}_{13} + \frac{1}{\sqrt{5}}\mathbf{e}_{14} + \frac{1}{\sqrt{5}}\mathbf{e}_{23} - \frac{1}{\sqrt{5}}\mathbf{e}_{24}$$

$$= \frac{1}{\sqrt{5}} u \wedge v$$

$$|w \wedge z| = 1.$$

4.4. $\mathbf{e}_{12} + 2\mathbf{e}_{13} + 2\mathbf{e}_{23} = (\mathbf{e}_1 + \mathbf{e}_2) \wedge (\mathbf{e}_2 + 2\mathbf{e}_3)$.

4.5. Method 1: Assume $\mathbf{e}_{12} + \mathbf{e}_{34} = (\Sigma a_i \mathbf{e}_i) \wedge (\Sigma b_j \mathbf{e}_j)$, and derive a contradiction.
Method 2: Clearly, if ξ is simple, $\xi \wedge \xi = 0$. Since $(\mathbf{e}_{12} + \mathbf{e}_{34}) \wedge (\mathbf{e}_{12} + \mathbf{e}_{34}) = 2\mathbf{e}_{1234} \neq 0$, it is not simple. (Actually, for $\xi \in \Lambda_2 \mathbf{R}^n$, ξ simple $\Leftrightarrow \xi \wedge \xi = 0$. For $\xi \in \Lambda_m \mathbf{R}^n$, $m > 2$, ξ simple $\Rightarrow \xi \wedge \xi = 0$.)

4.6. $\int_0^1 \int_0^1 \langle \mathbf{e}_{12}, \varphi \rangle \, dx_2 \, dx_1$, $\langle \mathbf{e}_{12}, \varphi \rangle = x_1 \sin x_1 x_2$.
Inside integral $= -\cos x_1 x_2]_{x_2=0}^{1} = 1 - \cos x_1$.
Outside integral $= x_1 - \sin x_1]_{x_1=0}^{1} = 1 - \sin 1$.

4.7. The surface is a unit disc with normal $\mathbf{e}_1 + \mathbf{e}_2 + \mathbf{e}_3$ unit tangent $\xi = (\mathbf{e}_{12} + \mathbf{e}_{23} - \mathbf{e}_{13})/\sqrt{3}$. $\varphi(\xi) = 4/\sqrt{3}$. Integral $= 4\pi/\sqrt{3}$.

4.8. If $I \in \mathbf{I}_m$, then $\partial I \in \mathscr{R}_{m-1}$ by definition and $\partial(\partial I) = 0 \in \mathscr{R}_{m-2}$. Therefore, $\partial I \in \mathbf{I}_{m-1}$. If $F \in \mathscr{F}_m$, then $F = T + \partial S$, with $T \in \mathscr{R}_m$, $S \in \mathscr{R}_{m+1}$. Since $\partial F = \partial T$, $\partial F \in \mathscr{F}_{m-1}$.

To prove that $\operatorname{spt} \partial T \subset \operatorname{spt} T$, consider a form $\varphi \in \mathscr{D}^{m-1}$ such that $\operatorname{spt} \varphi \cap \operatorname{spt} T = \varnothing$. Then $\operatorname{spt} d\varphi \cap \operatorname{spt} T = \varnothing$, and consequently $\partial T(\varphi) = T(d\varphi) = 0$. We conclude that $\operatorname{spt} \partial T \subset \operatorname{spt} T$.

4.9. (a)
$$T(f\,dx + g\,dy) = \int_0^1 f(x,0)\,dx.$$

$$\partial T(h) = T\left[\frac{\partial h}{\partial x}dx + \frac{\partial h}{\partial y}dy\right]$$

$$= \int_0^1 \frac{\partial h}{\partial x}(x,0)\,dx = h(1,0) - h(0,0).$$

Hence $\partial T = \mathscr{H}^0 \llcorner \{(1,0)\} - \mathscr{H}^0 \llcorner \{(0,0)\}$.

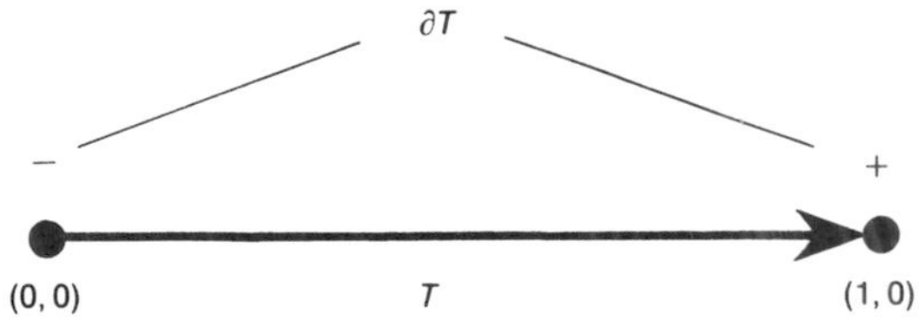

(b) Let $E = \{(x,x)\colon 0 \le x \le 1\}$.

$$T(f\,dx + g\,dy) = \int_E 3\sqrt{2}\,(f(x,x) + g(x,x))\,d\mathscr{H}^1$$

$$= \int_0^1 3\sqrt{2}\,(f(x,x) + g(x,x))\sqrt{2}\,dx$$

$$= 6\int_0^1 (f(x,x) + g(x,x))\,dx.$$

$$\partial T(h) = 6\int_0^1 \left[\frac{\partial x_1}{\partial h}(x,x) + \frac{\partial h}{\partial x_2}(x,x)\right]dx$$

$$= 6(h(1,1) - h(0,0)).$$

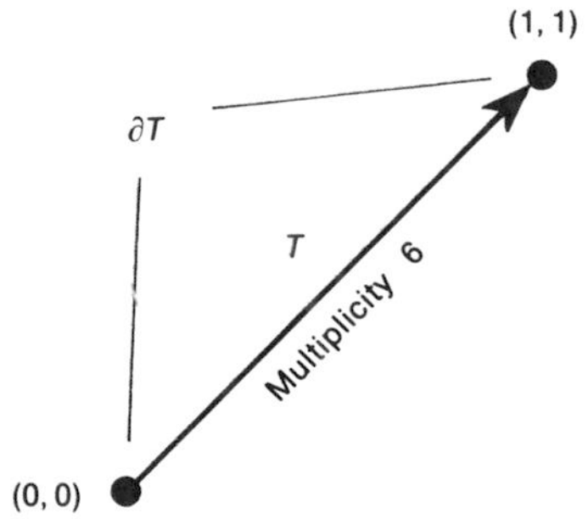

Notice that

$$T = 6\mathscr{H}^1 \llcorner E \wedge \frac{\mathbf{e}_1 + \mathbf{e}_2}{\sqrt{2}}.$$

4.10. It follows from Theorem 4.4(1) that $\mathbf{I}_m$ is $\mathbf{M}$ dense in $\mathscr{R}_m$. To show that $\mathbf{I}_m$ is $\mathscr{F}$ dense in $\mathscr{F}_m$, let $R \in \mathscr{F}_m$, so that $R = T + \partial S$, with $T \in \mathscr{R}_m$ and $S \in \mathscr{R}_{m+1}$. Given $\varepsilon > 0$, choose $T_1 \in \mathbf{I}_m$, $S_1 \in \mathbf{I}_{m+1}$, such that $\mathbf{M}(T_1 - T) + \mathbf{M}(S_1 - S) < \varepsilon$. Then $T_1 + \partial S_1 \in \mathbf{I}_m$, and $\mathscr{F}((T_1 + \partial S_1) - R) = \mathscr{F}((T_1 - T) + \partial(S_1 - S)) \leq \mathbf{M}(T_1 - T) + \mathbf{M}(S_1 - S) < \varepsilon$.

4.11. It follows from Theorem 4.4(1) that $A = \{T \in \mathscr{R}_m: \operatorname{spt} T \subset \mathbf{B}(\mathbf{0}, R)\}$ is $\mathbf{M}$ complete. To show that $B = \{T \in \mathscr{F}_m: \operatorname{spt} T \subset \mathbf{B}(\mathbf{0}, R)\}$ is $\mathscr{F}$-complete, let R_j be a Cauchy sequence in B. By taking a subsequence, we may assume that $\mathscr{F}(R_{j+1} - R_j) < 2^{-j}$. Write $R_{j+1} - R_j = T_j + \partial S_j$ with $\mathbf{M}(T_j) + \mathbf{M}(S_j) < 2^{-j}$. Since A is $\mathbf{M}$ complete, ΣT_j converges to a rectifiable current, T, and ΣS_j converges to a rectifiable current S. It is easy to check that $R_j \to R_1 + T + \partial S \in \mathscr{F}_m$.

4.12. That ∂ carries $\mathbf{N}_m$ into $\mathbf{N}_{m-1}$ follows immediately from the definition of $\mathbf{N}_m$ and the fact that $\partial\partial = 0$. Since ∂ is $\mathbf{F}$ continuous, it follows that ∂ carries $\mathbf{F}_m$ into $\mathbf{F}_{m-1}$.

4.13. The first follows immediately from the definitions. The second is the definition of $\mathbf{R}_m$.

4.14. The first follows immediately from the definitions. By Exercise 4.12, $\mathbf{F}_m \supset \{T + \partial S: T \in \mathbf{R}_m, S \in \mathbf{R}_{m+1}\}$. Conversely, suppose $R \in \mathbf{F}_m$. Let $N_j \in \mathbf{N}_m$ with $\mathbf{F}(N_j - R) < 2^{-j-1}$ and hence $\mathbf{F}(N_{j+1} - N_j) < 2^{-j}$. Hence $(N_{j+1} - N_j) = A_j + \partial B_j$ for currents A_j and B_j with $\mathbf{M}(A_j) + \mathbf{M}(B_j) < 2^{-j}$. Since $\mathbf{M}(\partial B_j) = \mathbf{M}(N_{j+1} - N_j - A_j) < \infty$, $B_j \in \mathbf{N}_{m+1}$. Therefore $A_j = N_{j+1} - N_j - \partial B_j \in \mathbf{N}_m$. By Proposition 4.6, $\Sigma A_j \in \mathbf{R}_m$ and $\Sigma B_j \in \mathbf{R}_{m+1}$. Finally, $R = (N_1 + \Sigma A_j) + \partial \Sigma B_j$ is of the form $T + \partial S$, as desired.

4.15. That $\mathbf{I}_m \subset \mathbf{N}_m$ follows immediately from the definitions. Next,

$$\mathscr{F}_m \subset \mathscr{F}\text{-closure of } \mathbf{I}_m \subset \mathbf{F}\text{-closure of } \mathbf{N}_m = \mathbf{F}_m;$$

the first inclusion follows from Exercise 4.10, whereas the second follows because $\mathbf{I}_m \subset \mathbf{N}_m$ and $\mathbf{F} \leq \mathscr{F}$. Finally $\mathscr{R}_m \subset \mathbf{R}_m$, because if $T \in \mathscr{R}_m$, then $T \in \mathscr{F}_m \subset \mathbf{F}_m$ and $\mathbf{M}(T) < \infty$.

4.16. (a)
$$T(f\,dx + g\,dy) = \sum_{k=1}^{\infty} \int_0^{2^{-k}} g(k^{-1}, y)\,dy.$$

$$\partial T(h) = \sum_{k=1}^{\infty} h(k^{-1}, 2^{-k}) - \sum_{k=1}^{\infty} h(k^{-1}, 0).$$

$$T \in \mathscr{P}_1.$$

(b)
$$T(f\,dx + g\,dy) = \int_0^1 \int_0^1 f(x, y)\,dx\,dy.$$

$$\partial T(h) = \int_0^1 h(1, y)\,dy - \int_0^1 h(0, y)\,dy.$$

$$T \in \mathbf{N}_1.$$

(c)
$$T(f\,dx + g\,dy) = \int_0^1 g(x,0)\,dx.$$
$$\partial T(h) = \int_0^1 \frac{\partial h}{\partial y}(x,0)\,dx.$$
$$T \in \mathscr{D}_1.$$

(See example following Theorem 4.7.)

(d)
$$T(f\,dx + g\,dy) = f(a).$$
$$\partial T(h) = \frac{\partial h}{\partial x}(a).$$
$$T \in \mathscr{D}_1.$$

(See Theorem 4.7.)

(e)
$$T(f\,dx + g\,dy) = \int_{\text{unit disc}} f(x,y)\,dx\,dy.$$
$$\partial T(h) = \int_0^1 h\left(\sqrt{1-y^2}, y\right) dy - \int_0^1 h\left(-\sqrt{1-y^2}, y\right) dy.$$
$$T \in \mathbf{N}_1.$$

4.17.
$$\mathscr{H}^1(E) = 1 - \sum_{n=1}^{\infty} 2^{n-1} \cdot 4^{-n} = 1/2.$$

Let T be the sum of all oriented intervals removed in defining E. Clearly $T \in \mathscr{R}_1 - \mathbf{I}_1$. Therefore $\mathscr{H}^1 \llcorner E \wedge \mathbf{i} = [0,1] - T \in \mathscr{R}_1 - \mathbf{I}_1$.

4.18.
$$\begin{aligned}\partial T &= (\partial T)\llcorner\{u > r\} + (\partial T)\llcorner\{u \le r\} \\ &= \partial\left[T\llcorner\{u > r\} + T\llcorner\{u \le r\}\right].\end{aligned}$$

4.19. Immediate from definition of $\langle T, u, r+\rangle$.

4.20.
$$\mathbf{M}\langle T, u, r+\rangle < \infty$$

for almost all r by (4).

$$\mathbf{M}(\partial\langle T, u, r+\rangle) = \mathbf{M}\langle \partial T, u, r+\rangle$$

(by (2))

$$< \infty$$

for almost all r by (4).

4.21. Choose currents A and B such that $T = A + \partial B$ and $\mathbf{F}(T) = \mathbf{M}(A) + \mathbf{M}(B)$. Since $\partial A = \partial T$, $A \in \mathbf{N}$. Since $\partial B = T - A$, $B \in \mathbf{N}$. Now

$$T\llcorner\{u \le r\} = A\llcorner\{u \le r\} + \partial\left[B\llcorner\{u \le r\}\right] - \langle B, u, r+\rangle$$

by 4.11(1). Therefore,

$$\mathbf{F}(T\llcorner\{u \le r\}) \le \mathbf{M}(A) + \mathbf{M}(B) + \mathbf{M}\langle B, u, r+\rangle.$$

Integration and 4.11(4) yield (6) as desired.

4.22. First consider the case $\mathbf{M}(T)<\infty$. Given $\varepsilon>0$, choose $\varphi\in\mathscr{D}^m$ with $\|\varphi(x)\|^*\le 1$ such that $\mathbf{M}(T)\le T(\varphi)+\varepsilon$. Then

$$\begin{aligned}\mathbf{M}(T)&\le T(\varphi)+\varepsilon=\lim T_i(\varphi)+\varepsilon\\&\le \liminf \mathbf{M}(T_i)+\varepsilon.\end{aligned}$$

Second, if $\mathbf{M}(T)=\infty$, given $\varepsilon>0$, choose $\varphi\in\mathscr{D}^m$ with $\|\varphi(x)\|^*\le 1$ such that $T(\varphi)>1/\varepsilon$. Then

$$\liminf \mathbf{M}(T_i)\ge \lim T_i(\varphi)>1/\varepsilon.$$

Hence, $\liminf \mathbf{M}(T_i)=\infty$, as desired.

4.23. We prove b, of which a is a special case.

$$\begin{aligned}f_\#S(\varphi)&=S(f^\#\varphi)=\int_E\langle\vec{S},f^\#\varphi\rangle l\,d\mathscr{H}^m\\&=\int_E\left\langle\Lambda_m(Df(x))(\vec{S}),\varphi(f(x))\right\rangle l(x)\,d\mathscr{H}^m x\\&=\int_E\left\langle\frac{\Lambda_m(Df(x))(\vec{S})}{|\Lambda_m(Df(x))(\vec{S})|},\varphi(f(x))\right\rangle l(x)\operatorname{ap}J_m(f|E)\,d\mathscr{H}^m x\end{aligned}$$

(where the contribution from points at which $|\Lambda_m(Df(x))(\vec{S})|=\operatorname{ap}J_m(f|E)=0$ is still interpreted to be 0)

$$=\int_{f(E)}\sum_{y=f(x)}\left\langle\frac{(\Lambda_m Df(x))(\vec{S})}{|(\Lambda_m Df(x))(\vec{S})|},\varphi\right\rangle l(x)\,d\mathscr{H}^m y$$

by the coarea formula, 3.13. Therefore,

$$f_\#S=(\mathscr{H}^m\llcorner f(E))\wedge\sum_{y=f(x)}l(x)\frac{(\Lambda_m Df(x))(\vec{S})}{|(\Lambda_m Df(x))(\vec{S})|}.$$

Chapter 5

5.1. Applying a homothety $\mu_r(x)=rx$ multiplies $\mathbf{M}(S)$ by r^{m+1}, $\mathbf{M}(T)$ by r^m, and hence both sides of the inequality by the same factor r^m.

5.2. Immediately from the definitions, $\mathbf{I}_{m+1}\subset\{T\in\mathscr{R}_{m+1}:\mathbf{M}(\partial T)<\infty\}$. The opposite inclusion follows from 5.4(1) because $\mathbf{I}_{m+1}$ is $\mathscr{F}$ dense in $\mathscr{F}_{m+1}$ (Exercise 4.10). Also from the definitions, $\mathscr{R}_m\subset\{T\in\mathscr{F}_m:\mathbf{M}(T)<\infty\}$. Conversely, suppose $T\in\mathscr{F}_m$ with $\mathbf{M}(T)<\infty$. Then $T=R+\partial S$, with $R\in\mathscr{R}_m$, $S\in\mathscr{R}_{m+1}$. Since $\mathbf{M}(\partial S)=\mathbf{M}(T-R)<\infty$, it follows from (2) that $S\in\mathbf{I}_{m+1}$. Therefore, $T=R+\partial S\in\mathscr{R}_m$, as desired.

5.3. One good candidate is the sequence

$$T_k=\sum_{j=1}^{2^k}\left[\left(j-\frac{1}{2}\right)2^{-k},j2^{-k}\right]\in\mathbf{I}_1\mathbf{R}^1,$$

which at first glance appears to converge to $\frac{1}{2}[0,1] \notin \mathbf{I}_1\mathbf{R}^1$.

$$\bullet\!\!\rightarrow \quad \rightarrow \quad \rightarrow \quad \rightarrow \quad \rightarrow \quad \rightarrow \quad \rightarrow \quad \rightarrow\!\!\bullet$$

$$0 \qquad\qquad\qquad\qquad T_3 \qquad\qquad\qquad\qquad 1$$

5.4. Let T be the unit circle $\mathbf{R}^2$ with multiplicity $1/N$. The the only normal current S with $S = \partial T$ is the unit disc with multiplicity $1/N$. Since $\mathbf{M}(S) = \pi/N$ and $\mathbf{M}(T) = 2\pi/N$, the isoperimetric inequality does not hold for any constant γ.

Chapter 6

6.1. Just plug $f(y,z) = y \tan z$ into the minimal surface equation, 6.1.

6.2. Just plug $f(x,y) = \ln(\cos x/\cos y)$ into the minimal surface equation, 6.1.

6.3. Apply the minimal surface equation to $z = f(x,y)$. Let $w = u + iv$. Then

$$f_x = \frac{\partial f/\partial u}{\partial x/\partial u} + \frac{\partial f/\partial v}{\partial x/\partial v},$$

etc.

6.4. A surface of revolution has an equation of the form $r = g(z)$, where $r = \sqrt{x^2 + y^2}$. Differentiating $g^2 = x^2 + y^2$ implicitly yields

$$gg'z_x = x, \quad gg'z_y = y,$$

$$\left(g'^2 + gg''\right)z_x^2 + gg'z_{xx} = 1,$$

$$\left(g'^2 + gg''\right)z_y^2 + gg'z_{yy} = 1,$$

$$\left(g'^2 + gg''\right)z_x z_y + gg'z_{xy} = 0.$$

Applying the minimal surface equation to $z(x,y)$ yields

$$\begin{aligned}
0 &= \left[\left(1 + z_y^2\right)z_{xx} + \left(1 + z_x^2\right)z_{yy} - 2z_x z_y z_{xy}\right] gg' \\
&= \left(1 - z_y^2\right)\left(1 - \left(g'^2 + gg''\right)z_x^2\right) \\
&\quad + \left(1 + z_x^2\right)\left(1 - \left(g'^2 + gg''\right)z_y^2\right) \\
&\quad + 2z_x z_y\left(g'^2 + gg''\right)z_x z_y \\
&= 2 + \left(z_x^2 + z_y^2\right)\left(1 - g'^2 - gg''\right) \\
&= \left(z_x^2 + z_y^2\right)\left(1 + g'^2 - gg''\right).
\end{aligned}$$

Therefore $gg'' = 1 + g'^2$. Substituting $p = g'$ yields $gp(dp/dg) = 1 + p^2$. Integration yields $p^2 = ag^2 - 1$, i.e.,

$$\frac{dg}{\sqrt{a^2g^2 - 1}} = \pm\, dz.$$

Integration yields $(1/a)\cosh^{-1} az = \pm z + c$, i.e., $r = g(z) = (1/a)\cosh(\pm az + c) = (1/a)\cosh(az \mp c)$, which is congruent to $r = (1/a)\cosh az$.

6.5. $$0 = \operatorname{div} \frac{\nabla f}{1 + |\nabla f|^2}$$

$$= \frac{\partial}{\partial x}\left[f_x\left(1 + f_x^2 + f_y^2\right)^{-1/2}\right] + \frac{\partial}{\partial y}\left[f_y\left(1 + f_x^2 + f_y^2\right)^{-1/2}\right]$$

$$= f_{xx}\left(1 + f_x^2 + f_y^2\right)^{-1/2} - f_x\left(1 + f_x^2 + f_y^2\right)^{-3/2}\left(f_x f_{xx} + f_y f_{xy}\right)$$

$$+ f_{yy}\left(1 + f_x^2 + f_y^2\right)^{-1/2} - f_y\left(1 + f_x^2 + f_y^2\right)^{-3/2}\left(f_x f_{xy} + f_y f_{yy}\right).$$

$$0 = \left(f_{xx} + f_{yy}\right)\left(1 + f_x^2 + f_y^2\right) - f_x^2 f_{xx} - 2 f_x f_y f_{xy} - f_y^2 f_{yy}$$

$$= \left(1 + f_y^2\right) f_{xx} - 2 f_x f_y f_{xy} + \left(1 + f_x^2\right) f_{yy}.$$

6.6. $$f(x, y) = \left(x^2 - y^2, 2xy\right).$$

$$\left(1 + |f_y|^2\right) f_{xx} - 2\left(f_x \cdot f_y\right) f_{xy} + \left(1 + |f_x|^2\right) f_{yy}$$

$$= \left(2 + 8|z|^2, 0\right) - 0$$

$$+ \left(-2 - 8|z|^2, 0\right)$$

$$= 0.$$

6.7. $g(z) = (u(x, y), v(x, y))$, satisfying the Cauchy–Riemann equations $u_x = v_y$, $u_y = -v_x$, and hence $u_{yy} = -u_{xx}$ and $v_{yy} = -v_{xx}$.

$$\left(1 + |f_y|^2\right) f_{xx} - 2\left(f_x \cdot f_y\right) f_{xy} + \left(1 + |f_x|^2\right) f_{yy}$$

$$= \left(1 + u_y^2 + v_y^2\right)\left(u_{xx}, v_{xx}\right) - 2\left(u_x u_y + v_x v_y\right)\left(u_{xy}, v_{xy}\right)$$

$$+ \left(1 + u_x^2 + v_x^2\right)\left(u_{yy}, v_{yy}\right)$$

$$= \left(1 + u_x^2 + u_y^2\right)\left(u_{xx}, v_{xx}\right) - 0 + \left(1 + u_x^2 + u_y^2\right)\left(-u_{xx}, -v_{xx}\right)$$

$$= 0$$

Chapter 8

8.1. To contradict 8.1, one might try putting a half-twist in the tail of Fig. 8.1(1), so that the surface, like 8.1(2), would not be orientable. However, there is another surface like 8.1(2), in which the little hole goes in underneath in front and comes out on top in back. To contradict 8.4, one might try a large horizontal disc centered at the origin and a little vertical disc tangent to it at the origin:

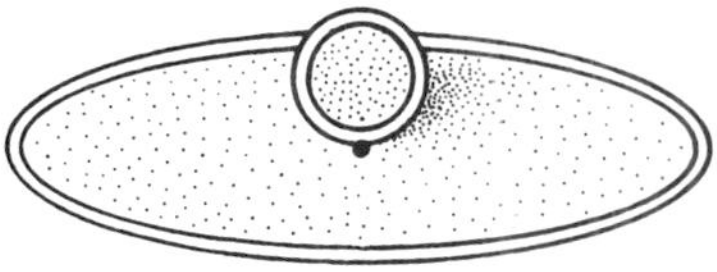

However, the area-minimizing surface is

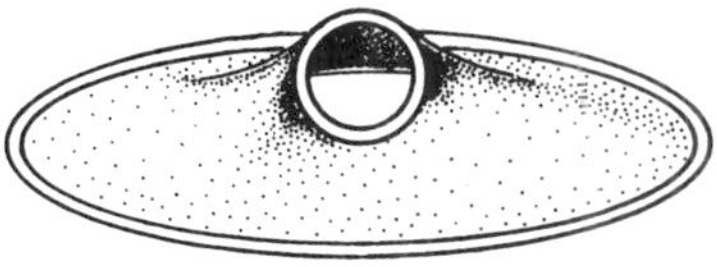

8.2.

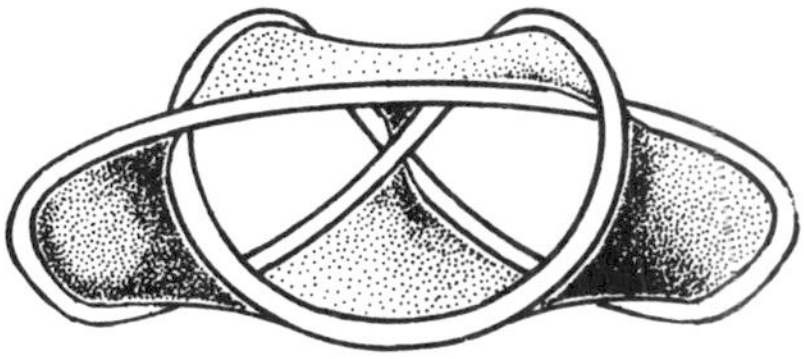

Front view

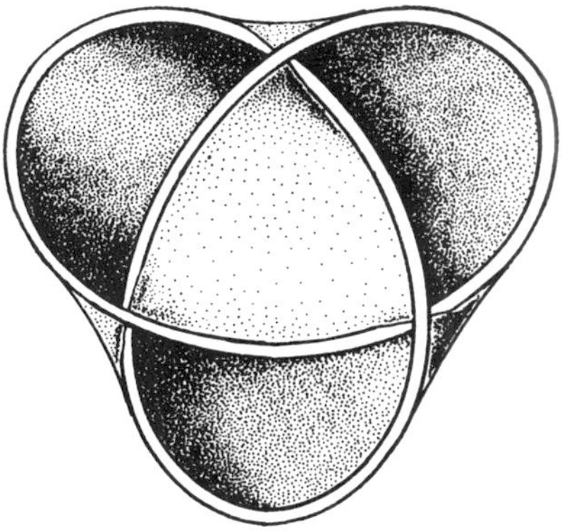

Top view

(There are two sheets in the middle.)

8.3. Let E be a unit element of surface area in $\mathbf{R}^4$ in a plane with orthonormal basis u, v. Let P, Q denote projection onto the x_1-x_2 and x_3-x_4 planes. Then

$$\begin{aligned}
\text{area } PE + \text{area } QE &= |Pu \wedge Pv| + |Qu \wedge Qv| \\
&\leq \left(|Pu|\,|Pv| + |Qu|\,|Qv| \right) \\
&\leq \left(|Pu|^2 + |Qu|^2 \right)^{1/2} \left(|Pv|^2 + |Qv|^2 \right)^{1/2} \\
&= |u|\,|v| = 1 = \text{area } E.
\end{aligned}$$

Therefore the area of any surface is at least the sum of the areas of its projections.

Now let S be a surface with the same boundary as the two discs $D_1 + D_2$. Since $\partial(P_\# S) = \partial D_1$ and $P_\# S$ and D_1 both lie in the x_1-x_2 plane, $P_\# S = D_1$. Similarly $Q_\# S = D_2$. Therefore

$$\begin{aligned}
\text{area } S &\geq \text{area } P_\# S + \text{area } Q_\# S \\
&= \text{area } D_1 + \text{area } D_2 \\
&= \text{area}(D_1 + D_2).
\end{aligned}$$

Therefore $D_1 + D_2$ is area minimizing.

Chapter 9

9.1. ∂R, where R is the pictured rectifiable current of infinitely many components C_j of length 2^{-j}.

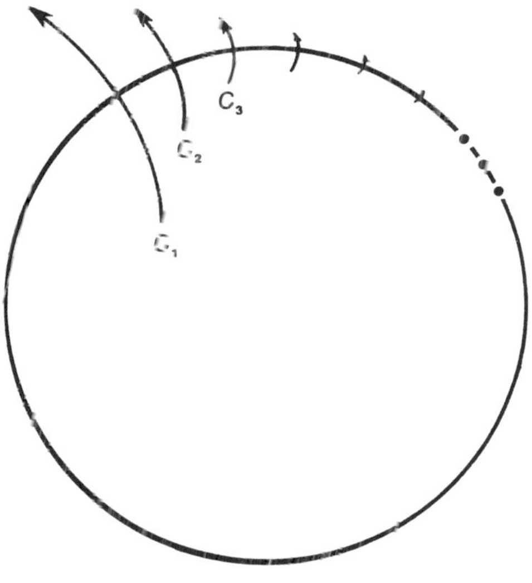

9.2. Suppose S is not area minimizing. For some $a, r > 0$ there is a rectifiable current T such that $\partial T = \partial(S \llcorner \mathbf{B}(a,r))$ and $\varepsilon = \mathbf{M}(S \llcorner \mathbf{B}(a,r)) - \mathbf{M}(T) > 0$. Choose j such that

$$\operatorname{spt}\left(S_j - S - (A + \partial B)\right) \cap \mathbf{U}(a, r+1) = \varnothing,$$

$A \in \mathscr{R}_m$, $B \in \mathscr{R}_{m+1}$, and $\mathbf{M}(A) + \mathbf{M}(B) < \varepsilon$. Let $u(x) = |x - a|$, and apply slicing theory, 4.11(4), to choose $r < s < r+1$ such that $\mathbf{M}\langle B, u, s+\rangle \leq \mathbf{M}(B)$. Now $S_j \llcorner \mathbf{B}(a,s)$ has the same boundary as

$$T + S \llcorner (\mathbf{B}(a,s) - \mathbf{B}(a,r)) + A \llcorner \mathbf{B}(a,s) - \langle B, u, s+\rangle$$

and more mass. Therefore S_j is not area minimizing.

9.3. Suppose $p \in (\operatorname{spt} S) \cap \{\sqrt{x^2 + y^2} < R - 2\sqrt{R}\}$. Then the distance from p to $\operatorname{spt} \partial S$ exceeds $2\sqrt{R}$. By monotonicity,

$$\mathbf{M}(S) > \pi\left(2\sqrt{R}\right)^2 = 4\pi R = \text{area cylinder},$$

which contradicts S area minimizing.

9.4. False:

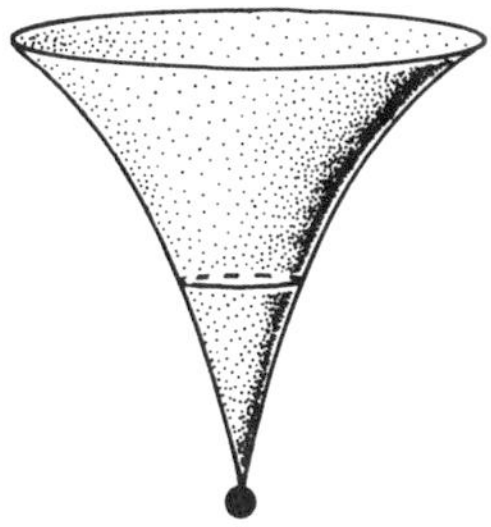

9.5. (a) One example is two unit orthogonal (complex) discs in $\mathbf{R}^4$, where the density jumps up to 2 at the origin (cf. 6.3 or Exercise 8.3).

(b) Suppose $x_i \to x$ but $f(x) < \overline{\lim} f(x_i)$. Choose $0 < r_0 < \operatorname{dist}(x, \operatorname{spt} \partial T)$ such that $\Theta^m(T, x, r_0) < \overline{\lim} f(x_i)$.

Choose $0 < r_1 < r_0$ such that

$$r_0^m \Theta^m(T, x, r_0) < r_i^m \overline{\lim} f(x_i).$$

Choose i such that $|x_i - x| < r_0 - r_1$ and

$$r_0^m \Theta^m(T, x, r_0) < r_1^m \Theta^m(T, x_i).$$

By monotonicity,

$$r_0^m \Theta^m(T, x, r_0) < r_1^m \Theta^m(T, x_i, r_1).$$

But since $\mathbf{B}(x, r_0) \supset \mathbf{B}(x_i, r_1)$,

$$r_0^m \Theta^m(T, x, r_0) \geq r_1^m \Theta^m(T, x_i, r_1).$$

This contradiction proves that f is upper semicontinuous.

9.6. Suppose $x \in \mathbf{S}(\mathbf{0}, 1) - \operatorname{Tan}(\operatorname{spt} T, \mathbf{0})$. Then for some $\varepsilon > 0$, for all sufficiently large r,

$$\operatorname{spt} \mu_{r\#} T \cap \mathbf{B}(x, \varepsilon) = \varnothing.$$

Consequently $x \notin \operatorname{spt} C$.

Figure 9.7.2 pictures an example in which $\operatorname{spt} C \neq \operatorname{Tan}(\operatorname{spt} T, \mathbf{0})$

9.7. $$C - D = \lim \mu_{r_j/s_j\#}\left[C - \mu_{r_j^{-1}\#} T\right] = 0.$$

9.8. (a) Let T_n be the homothetic expansion of T by 2^{n^2-n} (n odd). T_n consists of the interval $[2^{-n}, 2^{n-1}]$ on the x-axis, stuff outside $\mathbf{B}(\mathbf{0}, 2^{n-2})$, and stuff inside $\mathbf{B}(\mathbf{0}, 2^{-n})$ of total mass less than 2^{-n-2}. As a limit of the T_n, the nonnegative x-axis is an oriented tangent cone. Similarly taking n even yields the y-axis.

(b) Let S_n be the homothetic expansion of T by 2^{n^2} (n odd). S_n consists of the interval $[1, 2^{2n-1}]$ on the x-axis, the segment from $(0,1)$ to $(1,0)$, the interval $[2^{-2n-1}, 1]$ on the y-axis, stuff outside $\mathbf{B}(\mathbf{0}, 2^{2n-2})$, and stuff inside $\mathbf{B}(\mathbf{0}, 2^{-2n-1})$ of total mass less than 2^{-2n+1}. S_n converges to the interval $[1, \infty)$ on the x-axis, plus the segment from $(0,1)$ to $(1,0)$, plus the interval $[0,1]$ on the y-axis. This limit is not a cone.

(c) Looking at balls of radius 2^{-n^2} (n odd) exhibits the lower density 0. Looking at balls of radius 2^{-n^2} (n even) exhibits the upper density $1/2$. (Of course, for a subset of the nonnegative x-axis, the densities must lie between 0 and $1/2$.)

Chapter 10

10.1.

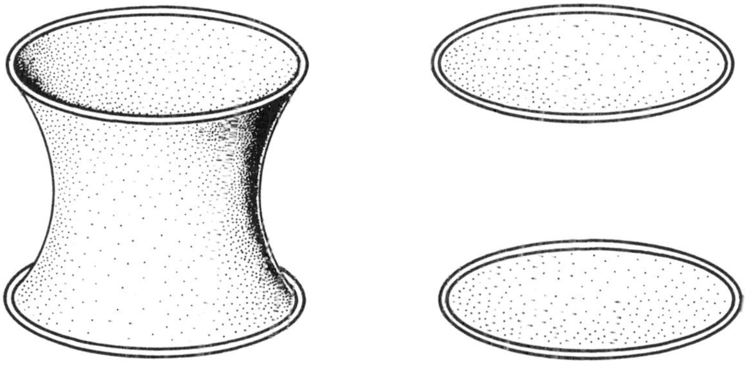

or see Figure 6.1.4

10.2. False. $[(-1,0,0),(1,0,0]+[(0,-1,100),(0,1,100)] \in \mathscr{R}_1 \mathbf{R}^2 \times \mathbf{R}^1$ is area minimizing, but its projection, $[(-1,0),(1,0)]+[(0,-1),(0,1)]$, is not.

10.3.

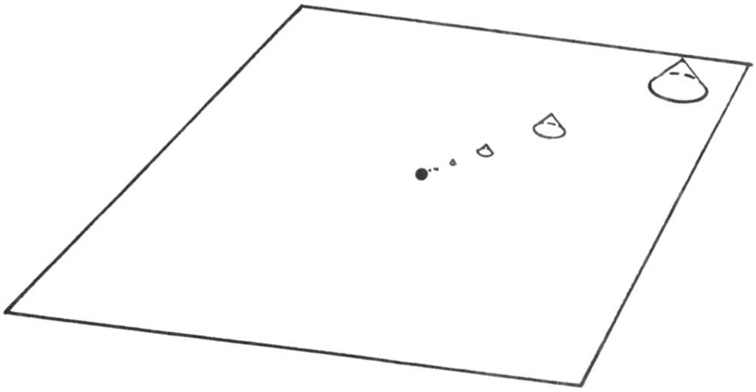

Chapter 11

11.1. Four similarly oriented, parallel circles

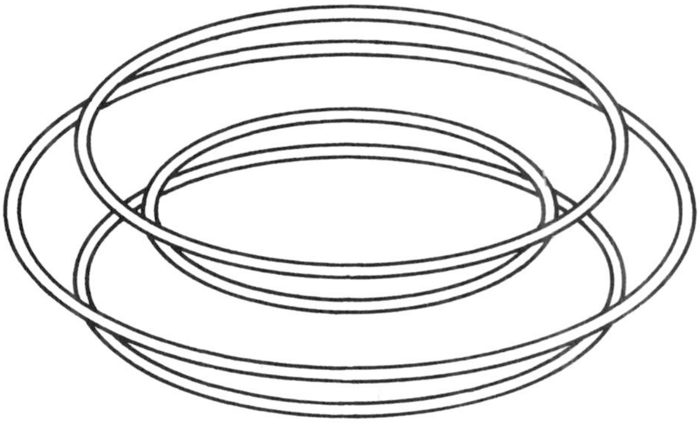

bound a catenoid and a horizontal annulus with cross-sections

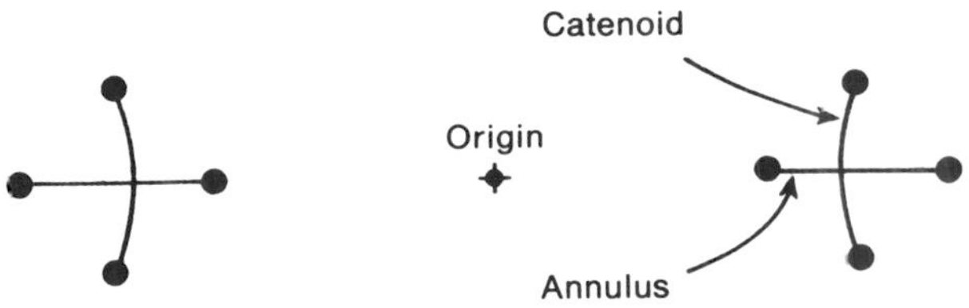

11.2. Because the tangent plane is constant, $\mathbf{v}(S)(\mathbf{R}^n \times G_2\mathbf{R}^n) = \text{area } S = \pi$.

11.3. Two close parallel unit discs, which can be deformed to a surface like that in Figure 11.1.2. Any unstable minimal surface, as in Figure 6.1.2, which can be deformed to a surface in Figure 6.1.4.

11.4. The surface of Figure 11.1.2. The catenoid has less area.

Bibliography

Section numbers follow the entries, indicating the locations in this book where the publications are cited.

Allard, William K. On the first variation of a varifold. *Ann. Math.* **95** (1972), 417–491. (§§5.3, 9.3, 10.2, 11.2).

Almgren, F. J., Jr.

[1] Existence and regularity almost everywhere of solutions to elliptic variational problems with constraints. *Mem. AMS No.* 165 (1976). (§§8.6, 11.3, 13.3, 13.4, 13.10)

[2] Optimal isoperimetric inequalities. *Indiana Univ. Math. J.* **35** (1986), 451–547. (§5.3)

[3] Q-valued functions minimizing Dirichlet's integral and the regularity of area minimizing rectifiable currents up to codimension two. *Bull. AMS* **8** (1983), 327–328. (§8.3)

[4] Questions and answers on area minimizing surfaces and geometric measure theory. *Proc. Symp. Pure Math.* **54** (1993), Part 1, 29–53. (Preface)

[5] Review of "Geometric Measure Theory: A Beginner's Guide." *Am. Math. Monthly* **96** (1989), 753–756. (Preface)

[6] Some interior regularity theorems for minimal surfaces and an extension of Bernstein's theorem. *Ann. Math.* **84** (1966), 277–292. (§8.1)

Almgren, F. J., Jr., and **Taylor**, J. E. Geometry of soap films. *Sci. Am.* **235** (1976), 82–93. (§§13.8, 13.9)

Bates, S. M. Toward a precise smoothness hypothesis in Sard's Theorem. *Proc. AMS* **117** (1993), 279–283. (§12.1)

Berger, M.

[1] "Geometry," Vol. II. Springer-Verlag, New York, 1972. (§13.2)

[2] Quelques problèmes de géométrie Riemannienne ou Deux variations sur les espaces symétriques compacts de rang un. *Enseignement Math.* **16** (1970), 73–96. (§6.5)

Besicovitch, A. S. A general form of the covering principle and relative differentiation of additive functions I, II. *Proc. Cambridge Phil. Soc.* **41** (1945), 103–110; **42** (1946), 1–10. (§2.7)

Bombieri, E., **De Giorgi**, E., and **Giusti**, E. Minimal cones and the Berstein problem. *Invent. Math.* **7** (1969), 243–268. (§§8.1, 10.7)

Brakke, Kenneth A.

[1] Minimal cones on hypercubes. *J. Geom. Anal.* **1** (1991), 329–338. (§6.5)

[2] Polyhedral minimal cones in $\mathbf{R}^4$. Preprint (1993). (§§6.5, 13.10)

[3] Soap films and covering spaces. Preprint (1993). (§6.5)

[4] The surface evolver. *Exp. Math.* **1** (1992), 141–165. (§§13.0, 13.12, 13.13)

[5] Century-old soap bubble problem solved! *Imagine That!* **3** (Fall, 1993), The Geometry Center, University of Minnesota, 1–3. (§13.13)

Brothers, J. E. (Ed.). Some open problems in geometric measure theory and its applications suggested by participants of the 1984 AMS Summer Institute. *In* "Geometric Measure Theory and the Calculus of Variations" (W. K. Allard and F. J. Almgren, Jr., Eds.). *Proc. Symp. Pure Math.* **44** (1986), 441–464. (Preface)

Brothers, John E., and **Morgan**, Frank. The isoperimetric theorem for general integrands. *Michigan Math. J.* **41** (1994), 419–431. (§13.2)

Burago, Yu. D., and **Zalgaller**, V. A. "Geometric Inequalities." Springer-Verlag, New York, 1988. (§13.2)

Chang, Sheldon. Two dimensional area minimizing integral currents are classical minimal surfaces. *J. AMS* **1** (1988), 699–778. (§8.3)

Costa, C. "Imersões Minimas Completas em $\mathbf{R}^3$ de Gênero um e Curvatura Total Finita." Doctoral thesis. IMPA, Rio de Janeiro, Brasil, 1982; Example of a complete minimal immersion in $\mathbf{R}^3$ of genus one and three embedded ends. *Bull. Soc. Bras. Mat.* **15** (1984), 47–54. (Fig. 6.1.3a)

Cox, Christopher, **Harrison**, Lisa, **Hutchings**, Michael, **Kim**, Susan, **Light**, Janette, **Mauer**, Andrew, and **Tilton**, Meg. "The Shortest Enclosure of Three Connected Areas in $\mathbf{R}^2$." NSF "SMALL" undergraduate research Geometry Group report. Williams College, Williamstown, MA, 1992. (§13.1)

Dao Trong Thi. Minimal real currents on compact Riemannian manifolds. *Izv. Akad. Nauk. SSSR Ser. Mat.* **41** (1977) [English translation in *Math. USSR Izv.* **11** (1977), 807–820]. (§6.5)

De Giorgi, E.

[1] "Frontiere Orientate di Misura Minima" (*Sem. Mat. Scuola Norm. Sup. Pisa*, 1960–1961). Editrice tecnico scientifica, Pisa, 1961. (§4.0)

[2] Sulla proprietà isoperimetrica dell'ipersfera, nella classe degli insiemi aventi frontiera orientata di misura finita. *Mem. Acc. Naz. Lincei*, Ser. 8, **5** (1958), 33–44. (§13.2)

[3] Su una teoria generale della misura $r-1$ dimensionale in un spazio ad r dimensioni. *Ann. Mat.* **4** (1955), 95–113. (§4.0)

de Rham, Georges.

[1] "On the Area of Complex Manifolds. Notes for the Seminar on Several Complex Variables." Institute for Advanced Study, Princeton, NJ, 1957–1958. (§6.5)

[2] Variétés différentiables, formes, courants, formes harmoniques. *Act. Sci. Indust.* **1222** (1955). (§4.0)

Douglas, Jesse. Solution of the problem of plateau. *Trans. AMS* **33** (1931), 263–321. (§1.2)

Falconer, K. J. "The Geometry of Fractal Sets." Cambridge Univ. Press, 1985. (§3.17)

Federer, Herbert.

[1] "**Geometric Measure Theory.**" Springer-Verlag, New York, 1969. (cited throughout)

[2] The singular sets of area minimizing rectifiable currents with codimension one and of area minimizing flat chains modulo two with arbitrary codimension. *Bull. AMS* **76** (1970), 767–771. (§§8.2, 11.1)

[3] Some theorems on integral currents. *Trans. AMS* **117** (1965), 43–67. (§6.5)

Federer, Herbert, and **Fleming**, Wendell H. Normal and integral currents. *Ann. Math.* **72** (1960), 458–520. (§§1.0, 4.0)

Fleming, Wendell H. On the oriented Plateau problem. *Rend. Circ. Mat. Palermo* (2)**11**, (1962), 1–22. (§8.1)

Foisy, Joel. "Soap Bubble Clusters in $\mathbf{R}^2$ and $\mathbf{R}^3$." Undergraduate thesis. Williams College, Williamstown, MA, 1991. (§13.0)

Foisy, Joel, **Alfaro**, Manuel, **Brock**, Jeffrey, **Hodges**, Nickelous, and **Zimba**, Jason. The standard double soap bubble in $\mathbf{R}^2$ uniquely minimizes perimeter. *Pac. J. Math.* **159** (1993), 47–59. Featured in the 1994 AMS "What's Happening in the Mathematical Sciences." (§13.1)

Fomenko, A. T. "The Plateau Problem. I: Historical Survey. II: Present State of the Theory." Gordon & Breach, New York, 1990. (Preface)

Giusti, Enrico. "Minimal Surfaces and Functions of Bounded Variation." Birkhäuser, Boston, 1984. (Pref., §12.4)

Gluck, Herman, **Mackenzie**, Dana, and **Morgan**, Frank. Volume-minimizing cycles in Grassmann manifolds. *Duke Math. J.*, to appear. (§6.5)

Gonzalez, E., **Massari**, U., and **Tamanini**, I. On the regularity of boundaries of sets minimizing perimeter with a volume constraint. *Indiana Univ. Math. J.* **32** (1983), 25–37. (§8.6)

Gromov, M. Isoperimetric inequalities in Riemannian manifolds, Appendix I to "Asymptotic Theory of Finite Dimensional Normed Spaces" by Vitali D.

Milman and Gideon Schechtman, Lecture Notes in Mathematics, No. 1200. Springer-Verlag, New York, 1986. (§13.2)

Grüter, Michael.

[1] Free boundaries in geometric measure theory and applications. *In* "Variational Methods for Free Surface Interfaces" (P. Concus and R. Finn, Eds.). Springer-Verlag, New York, 1986. (§12.3)

[2] Optimal regularity for codimension one minimal surfaces with a free boundary. *Manuscript Math.* **58** (1987), 295–343. (§12.3)

Hardt, Robert, and **Simon**, Leon.

[1] Boundary regularity and embedded solutions for the oriented Plateau problem. *Ann. Math.* **110** (1979), 439–486. (§8.4)

[2] "Seminar on Geometric Measure Theory." Birkhäuser, Boston, 1986. (Pref., §3.11)

Harrison, Jenny. Stokes' theorem for nonsmooth chains. *Bull. AMS* **29** (1993), 235–242. (§12.2)

Harvey, Reese, and **Lawson**, H. Blaine, Jr. Calibrated geometries. *Acta Math.* **148** (1982), 47–157. (§6.5)

Heppes, A. Isogonale sphärischen Netze. *Ann. Univ. Sci. Budapest Eötvös Sect. Math.* **7** (1964), 41–48. (§13.9)

Hildebrandt, Stefan. Free boundary problems for minimal surfaces and related questions. *Comm. Pure Appl. Math.* **39** (1986), $S111$–$S138$. (§12.3)

Hildebrandt, Stefan, and **Tromba**, Anthony. "Mathematics and Optimal Form." Scientific American Books (distributed by Freeman), New York, 1985. (Preface)

Hoffman, David. The computer-aided discovery of new embedded minimal surfaces. *Math. Intelligencer* **9** (1987), 8–21. (Fig. 6.1.3a)

Hoffman, David, and **Meeks**, W. H., III. A complete embedded minimal surface in $\mathbf{R}^3$ with genus one and three ends. *J. Diff. Geom.* **21** (1985), 109–127. (Fig. 6.1.3a)

Hoffman, David, **Wei**, Fusheng, and **Karcher**, Hermann. Adding handles to the helicoid. *Bull. AMS* **29** (1993), 77–84. (Pref., Fig. 6.1.3b)

Hopf, E. Elementare Bemerkungen über die Lösungen partieller Differentialgleichungen zweiter Ordnung vom elliptischen Typus, Sitzungberichte der Preussischen Akademie der Wissenshaften zu Berlin. *Phys.-Math. Klasse* **19** (1927), 147–152. (§§8.5, 10.4)

Hsiang, Wu-Yi. Talk at First MSJ International Research Institute, Sendai, 1993. (§13.2)

Hutchings, Michael. Connectivity in double soap bubbles. *J. Geom. Anal.*, to appear. (§13.0)

Kanigel, Robert. Bubble, bubble: Jean Taylor and the mathematics of minimal surfaces. *The Sciences* (May / June 1993), 32. (§13.8)

Kinderleharer, D., **Nirenberg**, L., and **Spruck**, J. Regularity in elliptic free boundary problems, I. *J. Anal. Math.* **34** (1978), 86–119. (§13.9)

Knorr, Wilbur Richard. "The Ancient Tradition of Geometric Problems." Birkhäuser, Boston (§13.2)

Lamarle, Ernest. Sur al stabilité des systèmes liquides en lames minces. *Mém. Acad. R. Belg.* **35** (1864), 3–104. (§13.9)

Lawlor, Gary

[1] The angle criterion. *Invent. Math.* **95** (1989), 437–446. (§6.5)

[2] A sufficient condition for a cone to be area-minimizing. *Mem. AMS* **91** (446) (1991). (§6.5)

Lawlor, Gary, and **Morgan**, Frank. Paired calibrations applied to soap films, immiscible fluids, and surfaces or networks minimizing other norms. *Pac. J. Math.* **166** (1994), 55–83. (§6.5)

Lawson, H. Blaine, Jr. "Lectures on Minimal Submanifolds," Vol. 1. Publish or Perish, 1980. (Pref., §§1.2, 8.1)

Lawson, H. Blaine, Jr., and **Osserman**, Robert. Non-existence, non-uniqueness and irregularity of solutions to the minimal surface system. *Acta Math.* **139** (1977), 1–17. (§6.2)

Le Hong Van. Relative calibrations and the problem of stability of minimal surfaces. *In* "Lecture Notes in Mathematics," No. 1453, pp. 245–262. Springer-Verlag, New York, 1990. (§6.5)

Mackenzie, D. See Nance.

Mandelbrot, Benoit B.

[1] "The Fractal Geometry of Nature." Freeman, New York, 1983. (§2.3)

[2] "Fractals." Freeman, San Francisco, 1977. (§2.3)

Morgan, Frank

[1] Area-minimizing surfaces, faces of Grassmannians, and calibrations. *Am. Math. Monthly* **95** (1988), 813–822. (§§4.1, 6.5)

[2] Calibrations and new singularities in area-minimizing surfaces: a survey. *In* "Variational Methods" (Proc. Conf. Paris, June 1988), (H. Berestycki, J.-M. Coron, and I. Ekeland, Eds.). *Prog. Nonlinear Diff. Eqns. Applns.* **4**, 329–342. Birkhäuser, Boston, 1990. (§6.5)

[3] Clusters minimizing area plus length of singular curves. *Math. Ann.* **299** (1994), 697–714. (§§13.0, 13.3, 13.11)

[4] "Compound Soap Bubbles, Shortest Networks, and Minimal Surfaces." AMS video, 1993. (§13.12)

[5] On finiteness of the number of stable minimal hypersurfaces with a fixed boundary. *Indiana Univ. Math. J.* **35** (1986), 779–833. (§6.1)

[6] Generic uniqueness results for hypersurfaces minimizing the integral of an elliptic integrand with constant coefficients. *Indiana Univ. Math. J.* **30** (1981), 29–45. (§8.4)

[7] (M, ε, δ)-minimal curve regularity. *Proc. AMS* **120** (1994), 677–686. (§§11.3, 13.10)

[8] Mathematicians, including undergraduates, look at soap bubbles. *Am. Math. Monthly* **101** (1994), 343–351. (§13.12)

[9] Minimal surfaces, crystals, shortest networks, and undergraduate research. *Math. Intelligencer* **14** (Summer 1992), 37–44. (§13.12)

[10] Review of "Mathematics and Optimal Form" by S. Hildebrandt and A. Tromba. *Am. Math. Monthly* **95** (1988), 569–575. (Preface)

[11] "Riemannian Geometry: A Beginner's Guide." A. K. Peters, Wellesley 1993. (§13.2)

[12] Size-minimizing rectifiable currents. *Invent. Math.* **96** (1989), 333–348. (§11.3)

[13] Soap bubbles and soap films. *In* "Mathematical Vistas: New and Recent Publications in Mathematics from the New York Academy of Sciences" (J. Malkevitch and D. McCarthy, Eds.), Vol. 607. New York Academy of Sciences, New York, 1990. (§6.1)

[14] Soap bubbles in $\mathbf{R}^2$ and in surfaces. *Pac. J. Math.* **165** (1994), 141–155. (§§13.1, 13.4, 13.10)

[15] Soap films and mathematics. *Proc. Symp. Pure Math.* **54** (1993), Part 1, 375–380. (§11.3)

[16] Soap films and problems without unique solutions. *Am. Sci.* **74** (1986), 232–236. (§6.1)

[17] Surfaces minimizing area plus length of singular curves. *Proc. AMS* **122** (1994), 1153–1164. (§13.11)

[18] Survey lectures on geometric measure theory. *In* "Geometry and Global Analysis: Report of the First MSJ International Research Institute, July 12–13, 1993," (Takeshi Kotake, Seiki Nishikawa, and Richard Schoen, Eds.), 87–110. Tohoku Univ., Math. Inst., Sendai, Japan, 1993. (§11.3)

Morgan, Frank, and **Taylor**, Jean. The tetrahedral point junction is excluded if triple junctions have edge energy. *Scr. Metall. Mater.* **15** (1991), 1907–1910. (§13.11)

Murdoch, Timothy A. Twisted calibrations. *Trans. AMS* **328** (1991), 239–257. (§6.5)

Nance [Mackenzie], Dana. Sufficient conditions for a pair of n-planes to be area-minimizing. *Math. Ann.* **279** (1987), 161–164. (§6.5)

Nitsche, Johannes C. C.

[1] The higher regularity of liquid edges in aggregates of minimal surfaces. *Nachr. Akad. Wiss. Göttingen Math.-Phys. Klasse* **2** (1977), 75–95. (§13.9)

[2] "Vorlesungen über Minimalflächen." Springer-Verlag, New York, 1975 [Translation: "Lectures on Minimal Surfaces." Cambridge Univ. Press, New York, 1989]. (Pref., §§1.2, 8.1, 8.4)

Osserman, Robert. "A Survey of Minimal Surfaces." New Dover edition, New York, 1986. (Pref., §§1.2, 8.1)

Plateau, J. A. F. "Statique Experimentale et Theorique des Liquides Soumis aux Seules Forces Moleculaires," Gauthier-Villars, Paris, 1873. (Pref., §13.8)

Rado, Tibor. "On the Problem of Plateau." Springer-Verlag, New York, 1933 [reprinted 1971]. (§1.2)

Reifenberg, E. R.

[1] Solution of the Plateau problem for m-dimensional surfaces of varying topological type. *Acta Math.* **104** (1960), 1–92. (§§4.0, 13.9)

[2] An epiperimetric inequality related to the analyticity of minimal surfaces. *Ann. Math.* **80** (1964), 1–14. (§§4.0, 13.9)

[3] On the analyticity of minimal surfaces. *Ann. Math.* **80** (1964), 15–21. (§§4.0, 13.9)

[4] A problem on circles. *Math. Gazette* **32** (1948), 290–292. (§2.7)

Ritoré, Manuel, and **Ros**, Antonio. Stable constant mean curvature tori and the isoperimetric problem in three space forms. *Comm. Math. Helv.* **67** (1992), 293–305. (§13.2)

Schoen, R., and **Simon**, L. Regularity of stable minimal hypersurfaces. *Comm. Pure Appl. Math.* **34** (1981), 741–797. (§8.5)

Schoen, R., **Simon**, L., and **Almgren**, F. J. Regularity and singularity estimates on hypersurfaces minimizing parametric elliptic variational integrals. *Acta Math.* **139** (1977), 217–265. (§8.5)

Schoen, R., **Simon**, L., and **Yau**, S.-T. Curvature estimates for minimal hypersurfaces. *Acta Math.* **134** (1975), 275–288. (§8.5)

Schoen, R., and **Yau**, S.-T. On the proof of the positive mass conjecture in General Relativity. *Comm. Math. Phys.* **65** (1979), 45–76. (Preface)

Schwarz, H. A. Beweis des Satzes, dass die Kugel kleinere Oberfläche besitzt, als jeder andere Körper gleichen Volumnes. *Nach. Ges. Wiss. Göttingen, Geschaeftliche, Mitt.* (1884), 1–13. (§13.2)

Serrin, J. On the strong maximum principle for quasilinear second order differential inequalities. *J. Funct. Anal.* **5** (1970), 184–193. (§8.5)

Simon, Leon

[1] Cylindrical tangent cones and the singular set of minimal submanifolds. *J. Diff. Geom.* **38** (1993), 585–652. (§11.1)

[2] Lectures on geometric measure theory. *Proc. Centre Math. Anal. Austral. Nat. Univ.* **3** (1983). (Pref., §§3.11, 4.3, 4.11, 9.1, 12.4)

[3] Survey lectures on minimal submanifolds. *In* "Seminar on Minimal Submanifolds" (E. Bombieri, Ed.), *Ann. of Math. Studies* **103**, Princeton Univ. Press, Princeton, NJ, 1983. (Preface)

Simons, James. Minimal varieties in Riemannian manifolds. *Ann. Math.* **88** (1968), 62–105. (§§8.1, 10.5)

Sommerville, D. M. Y. "An Introduction to the Geometry of n Dimensions." Dutton, New York, 1929. (§4.1)

Sullivan, John M. Sphere packings give an explicit bound for the Besicovitch covering theorem. *J. Geom. Anal.* **4** (1994), 219–231. (Fig. 2.7.1)

Taylor, Jean E.

[1] Boundary regularity for solutions to various capillarity and free boundary problems. *Com. PDE* **2** (1977), 323–357. (§12.3)

[2] Crystalline variational problems. *Bull. AMS* **84** (1987), 568–588. (§12.5)

[3] Regularity of the singular sets of two-dimensional area-minimizing flat chains modulo 3 in $\mathbf{R}^3$. *Invent. Math.* **22** (1973), 119–159. (§11.1)

[4] The structure of singularities in soap-bubble-like and soap-film-like minimal surfaces. *Ann. Math.* **103** (1976), 489–539. (§§11.3, 13.9)

Taylor, J. E., **Cahn**, J. W., and **Handwerker**, C. A. Geometric models of crystal growth. *Acta Metall. Mater.* **40** (1992), 1443–1474 (Overview No. 98–I). (§13.12)

Weaire, Denis. Froths, foams and heady geometry. *New Scientist*, (May 21, 1994). (§13.13)

White, Brian

[1] Existence of least-area mappings of N-dimensional domains. *Ann. Math.* **118** (1983), 179–185. (§1.2)

[2] A new proof of the compactness theorem for integral currents. *Comm. Math. Helv.* **64** (1989), 207–220. (§§3.17, 4.12, 5.4)

[3] Regularity of area-minimizing hypersurfaces at boundaries with multiplicity. *In* "Seminar on Minimal Submanifolds" (E. Bombieri, Ed.), pp. 293–301. Princeton Univ. Press, Princeton, NJ, 1983. (§8.4)

[4] A regularity theorem for minimizing hypersurfaces modulo p. *Proc. Symp. Pure Math.* **44** (1986), 413–427. (§§8.5, 11.1, 13.10)

[5] The structure of minimizing hypersurfaces mod 4. *Invent. Math.* **53** (1979), 45–58. (§11.1)

[6] "Regularity of the Singular Sets for Plateau-type Problems." In preparation.

Whitney, Hassler. "Geometric Integration Theory." Princeton Univ. Press, Princeton, NJ, 1957. (§4.7)

Wirtinger, W. Eine Determinantenidentität und ihre Anwendung auf analytische Gebilde und Hermitesche Massbestimmung. *Monatsh. Math. Phys.* **44** (1936), 343–365. (§6.5)

Wong, Yung-Chow. Differential geometry of Grassmann manifolds. *Proc. Natl. Acad. Sci. USA* **57** (1967), 589–594. (§4.1)

Young, L. C.

[1] On generalized surfaces of finite topological types. *Mem. AMS* No. 17, (1955), 1–63. (§4.0)

[2] Surfaces paramétriques generalisées. *Bull. Soc. Math. France* **79** (1951), 59–84. (§4.0)

Zworski, Maciej. Decomposition of normal currents. *Proc. AMS* **102** (1988), 831–839. (§4.5)

Index of Symbols

Name Index

(See also the bibliography, which includes cross-references to each citation within this text.)

Subject Index

On my way.

Photograph courtesy of the Morgan family;
taken by the author's grandfather, Dr. Charles W. Selemeyer.